Model-Driven
Software Engineering in Practice

Second Edition

Synthesis Lectures on Software Engineering

Editor
Luciano Baresi, *Politecnico di Milano*

The Synthesis Lectures on Software Engineering series publishes short books (75-125 pages) on conceiving, specifying, architecting, designing, implementing, managing, measuring, analyzing, validating, and verifying complex software systems. The goal is to provide both focused monographs on the different phases of the software process and detailed presentations of frontier topics. Premier software engineering conferences, such as ICSE, ESEC/FSE, and ASE will help shape the purview of the series and make it evolve.

Model-Driven Software Engineering in Practice: Second Edition

Marco Brambilla, Jordi Cabot, and Manuel Wimmer

ISBN: 978-3-031-01421-5   paperback
ISBN: 978-3-031-02549-5   ebook
ISBN: 978-3-031-03677-4   epub

DOI 10.1007/978-3-031-02549-5

A Publication in the Springer series
SYNTHESIS LECTURES ON SOFTWARE ENGINEERING

Lecture #4
Series Editor: Luciano Baresi, *Politecnico di Milano*
Series ISSN
Print 2328-3319   Electronic 2328-3327

Model-Driven
Software Engineering in Practice

Second Edition

Marco Brambilla
Politecnico di Milano, Italy

Jordi Cabot
ICREA and Open University of Catalonia (UOC), Spain

Manuel Wimmer
TU Wien, Austria

SYNTHESIS LECTURES ON SOFTWARE ENGINEERING #4

ABSTRACT

This book discusses how model-based approaches can improve the daily practice of software professionals. This is known as Model-Driven Software Engineering (MDSE) or, simply, Model-Driven Engineering (MDE).

MDSE practices have proved to increase efficiency and effectiveness in software development, as demonstrated by various quantitative and qualitative studies. MDSE adoption in the software industry is foreseen to grow exponentially in the near future, e.g., due to the convergence of software development and business analysis.

The aim of this book is to provide you with an agile and flexible tool to introduce you to the MDSE world, thus allowing you to quickly understand its basic principles and techniques and to choose the right set of MDSE instruments for your needs so that you can start to benefit from MDSE right away.

The book is organized into two main parts.

- The first part discusses the *foundations of MDSE* in terms of basic concepts (i.e., models and transformations), driving principles, application scenarios, and current standards, like the well-known MDA initiative proposed by OMG (Object Management Group) as well as the practices on how to integrate MDSE in existing development processes.

- The second part deals with the *technical aspects of MDSE*, spanning from the basics on when and how to build a domain-specific modeling language, to the description of Model-to-Text and Model-to-Model transformations, and the tools that support the management of MDSE projects.

The second edition of the book features:

- a set of completely new topics, including: full example of the creation of a new modeling language (IFML), discussion of modeling issues and approaches in specific domains, like business process modeling, user interaction modeling, and enterprise architecture

- complete revision of examples, figures, and text, for improving readability, understandability, and coherence

- better formulation of definitions, dependencies between concepts and ideas

- addition of a complete index of book content

KEYWORDS

modeling, software engineering, UML, domain-specific language, model-driven engineering, code generation, reverse engineering, model transformation, MDD, MDA, MDE, MDSE, OMG, DSL, EMF, Eclipse

Contents

Foreword

Technology takes forever to transition from academia to industry. At least it seems like forever. I had the honor to work with some of the original Multics operating system development team in the 1970s (some of them had been at it since the early 1960s). It seems almost comical to point out that Honeywell only ever sold a few dozen Multics mainframes, but they were advanced, really advanced—many of Multics' innovations (segmented memory, hardware security and privacy, multi-level security, etc.) took literally decades to find their way into other commercial products. I have a very distinct memory of looking at the original Intel 386 chip, impressed that the engineers had finally put Multics-style ring security right in the hardware, and less impressed when I discovered that they had done it exactly backward, with highly secure users unable to access low-security areas, but low-security users able to access the kernel. Technology transfer is a difficult and delicate task!

When I had the opportunity to help introduce a new technology and manage hype around that technology, I took it. At loose ends in 1989, I agreed to join the founding team of the Object Management Group (OMG), to help define commercial uptake for Object Technology (called object-oriented systems in the academic world, at least since Simula in 1967), and equally to help control the hype around the Object Technology marketplace. Having participated in the Artificial Intelligence (AI, or expert systems) world in the 1980s, I really didn't want to see another market meltdown as we'd experienced in AI: from the cover of *Time* magazine to a dead market in only five years!

That worked. OMG named, and helped define, the middleware marketplace that flourished in the 1990s, and continues today. Middleware ranges from: TCP socket-based, hand-defined protocols (generally an awful idea); to object-based, request-broker style stacks with automatically defined protocols from interface specifications (like OMG's own CORBA); to similarly automatically-defined, but publish-and-subscribe based protocols (like OMG's own DDS); to semantic integrate middleware with high-end built-in inference engines; to commercial everything-but-the-kitchen-sink "enterprise service bus" collections of request brokers, publish-and-subscribe, expert-system based, automatic-routing, voice-and-audio streaming lollapaloozas. Middleware abounds, and although there's still innovation, it's a very mature marketplace.

By the late 1990s, it was clear that the rapid rise of standardization focused on vertical markets (like healthcare IT, telecommunications, manufacturing, and financial services, OMG's initial range of so-called "domain" standards) would need something stronger than interface definition languages; to be more useful, standards in vertical markets (and arguably, all standards) should be defined using high-level, precise but abstract "modeling" languages. This class of languages should be much closer to user requirements, more readable by non-technical people, more focused on capturing process and semantics; in general, they should be more expressive. The natural choice for OMG and its members was of course OMG's own Unified Modeling Language (UML), standardized in 1997 as an experimental use of OMG's standards process that had heretofore focused on middleware. Even better, the UML standardization effort had produced a little-known but critical modeling language called the Meta-Object Facility (MOF) for defining modeling languages. This core of MOF plus extensible, profileable UML would easily be the foundation for a revolution in software development—and beyond.

As the millennium approached, OMG's senior staff met to consider how we could nudge the OMG membership in this valuable new direction. We came up with a name (Model-Driven Architecture, or MDA); a picture (which hasn't been universally adopted, but still helped make the transition); and a well-received white paper that explained why MDA would be the next logical step in the evolution of software engineering (and by extension, how it matches modeling in other engineering disciplines, though generally with other names, like "blueprints.") OMG's senior staff then spent a few months pitching this idea to our leading members, to a very mixed review. Some had been thinking this way for years and welcomed the approach; while some thought that it would be seen as an abandonment of our traditional middleware space (which by the way, we have never abandoned; the latest OMG middleware standards are weeks old at this writing and many more are to come). The CEO of one of our key member companies found the concept laughable, and in a memorable phrase, asked "Where's the sparkle?"

I truly believe, however, that organizations which resist change are the least stable. OMG therefore carried on and in 2001 introduced Model-Driven Architecture to a waiting world with a series of one-day events around the world. David Frankel's eponymous book, written and edited as he and I flew around the world to introduce the MDA concept, came out shortly thereafter; key influencers joined us in the campaign to add major new OMG standardization efforts in the modeling space. We would continue to create, extend, and support existing standards and new standards in the middleware and vertical-market spaces, but we would add significant new activities. It occurred to me that we actually had already been in the modeling space from the beginning; one can think of the Interface Definition Language of CORBA and DDS as simply a poor-man's modeling language, with limited expression of semantics.

For a while, the "sparkle" our members looked for was in academia primarily. As an avid participant in several academic conferences a year, I can tell you that uptake of MDA concepts (and terminology, like "platform-specific model" and "platform-independent model") took off like a rocket in universities. It took time, but the next step was a technology "pull" from engineering organizations that needed to perform better than the past (many of whom had already been using MDA techniques, and now had a name to point to); the creation of the Eclipse Foundation, starting in 2002, and its early embrace of modeling technology, also helped greatly. By 2010, modeling was firmly embedded in the world's software engineering psyche, and Gartner and Forrester were reporting that more than 71 UML tools were available on the market and adopted at some level. That's some serious "sparkle," and OMG members reveled in the success.

An interesting parallel world began to appear around MOF and UML, recognizing that modeling languages didn't have to be limited to modeling software systems (or "software intensive systems," as many called them); that, in fact, most complex software systems have to interact with other complex engineered systems, from building architecture to complex devices like mobile phones and aircraft carriers. We decided to roll out an entire fleet of MOF-defined languages to address the needs of many different modelers and marketplaces:

- UML System on a Chip: for microchip hardware/firmware/software definition;

- SoaML: for service-oriented architectures;

- BPMN: for business process modelers;

- BMM: for modeling the motivations and structure of a business;

- SysML: for modeling large, complex systems of software, hardware, facilities, people, and processes;

- UPDM: for modeling enterprise architectures;

- CWM: for data warehouses.

Each of these have been successful in a well-defined marketplace, often replacing a mix of many other languages and techniques that have fragmented a market and market opportunity. Along the way, our terminology morphed, changed, and extended, with arguments about the difference between "model-driven" and "model-based;" one of my favorite memories is of a keynote speech I gave just a couple of years ago in Oslo, after which an attendee came up to argue with me about my definition of the phrase "model-driven architecture." He wasn't particularly impressed that I had made up the term; it reminded me of a classic (and possibly apocryphal) story about the brilliant pianist Glenn Gould, who when accosted by a composer for his interpretation of the composer's work, yelled, "You don't understand your own composition!"

Over the past decade many new phrases have appeared around MDA, and one of the ones I consider most apt is Model-Driven Software Engineering (MDSE). This history lesson brings us to the work of this book, to help the neophyte understand and succeed with the technologies that make up MDSE. What are these mystical "modeling languages," how do we transform (compile) from one to another, and most importantly, how does this approach bring down the cost of development, maintenance, and integration of these systems? These aren't mysteries at all, and this book does a great job enlightening us on the techniques to get the most from a model-driven approach.

I'd like to leave you, dear reader, with one more thought. Recently, I had the opportunity to create, with dear friends Ivar Jacobson and Bertrand Meyer, an international community dedicated to better formalizing the software development process, and moving software development out of the fragmented "stone age" of insufficient theory chasing overwhelming need, to the ordered, structured engineering world on which other engineering disciplines depend. The Software Engineering Method and Theory (Semat) project brings together like-minded people worldwide to help bring software development into the 21st century, much as building architecture was driven into modernism by growing size and failures a millennium ago, and the shipbuilding industry had to formalize into ship blueprints some four centuries ago. My dream is that software engineering becomes engineering, and the huge stack of should-be-integrated engineering disciplines (civil, materials, software, hardware, etc.) be integrated into Model-Driven Engineering.

In your hands is part of the first step.

Richard Mark Soley, Ph.D.
Chairman and Chief Executive Officer
Object Management Group, Inc.
June 2012

10,000 meters over the central United States

Acknowledgments

This book wouldn't be the same without all the enriching discussions we have had with many other MDSE fans (and detractors!) during the last years—in person or within online forums. It would be almost impossible to list all of them here and therefore we wish to thank them all and to acknowledge their direct or indirect contribution to this book and to the MDE field at large, especially our current and previous colleagues.

An explicit mention must go to the ones who concretely helped us in the writing of this book. First of all, thanks to Diane Cerra, our Managing Editor at Morgan & Claypool, who believed in our project since the beginning and followed us with infinite patience throughout the whole book production process.

Secondly, thanks to Richard Soley, Chairman and CEO of OMG, who graciously agreed to introduce our work with his authoritative foreword.

And finally, last but not least, thanks to all the people that helped review the book: Ed Seidewitz (Model Driven Solutions), Davide di Ruscio (L'Aquila University), Juan Carlos Molina (Integranova), Vicente Pelechano (Polytechnic University of Valencia), and a bunch of our own colleagues and friends who carefully read and commented on what we were writing.

Marco Brambilla, Jordi Cabot, and Manuel Wimmer
January 2017

CHAPTER 1

Introduction

The human mind inadvertently and continuously re-works reality by applying cognitive processes that alter the subjective perception of it. Among the various cognitive processes that are applied, abstraction is one of the most prominent ones. In simple words, abstraction consists of the capability of finding the commonality in many different observations and thus generating a mental representation of the reality which is at the same time able to:

- generalize specific features of real objects (generalization);

- classify the objects into coherent clusters (classification); and

- aggregate objects into more complex ones (aggregation).

Actually, generalization, classification, and aggregation represent natural behaviors that the human mind is natively able to perform (babies start performing them since they are a few months old) and that are performed by people in their everyday life. Abstraction is also widely applied in science and technology, where it is often referred to as *modeling*. We can informally define a model as a simplified or partial representation of reality, defined in order to accomplish a task or to reach an agreement on a topic. Therefore, by definition, a model will never describe reality in its entirety.

1.1 PURPOSE AND USE OF MODELS

Models have been, and are, of central importance in many scientific contexts. Just think about physics or chemistry: the billiard ball model of a gas or the Bohr model of the atom are probably unacceptable simplifications of reality from many points of view, but at the same time have been paramount for understanding the basics of these fields; the uniform motion model in physics is something that will never be accomplished in the real world, but is extremely useful for teaching purposes and as a basis for subsequent, more complex theories. Mathematics and other formal descriptions have been extremely useful in all fields for modeling and building upon models. This has been proven very effective at description and powerful at prediction.

A huge branch of philosophy of science itself is based on models. Thinking about models at the abstract and philosophical level raises questions in semantics (i.e., the representational function performed by models), ontology (i.e., the kind of things that models are), epistemology (i.e., how to learn through or from models), and philosophy.

Models are recognized to implement at least two roles by applying abstraction:

- *reduction* feature: models only reflect a (relevant) selection of the original's properties, so as to focus on the aspects of interest; and

- *mapping* feature: models are based on an original individual, which is taken as a prototype of a category of individual and is abstracted and generalized to a model.

The *purpose* of models can be different too: they can be used for *descriptive purposes* (i.e., for describing the reality of a system or a context), *prescriptive purposes* (i.e., for determining the scope and details at which to study a problem), or for defining how a system shall be implemented.

In many senses, also considering that it is recognized that observer and observations alter the reality itself, at a philosophical level one can agree that "everything is a model," since nothing can be processed by the human mind without being "modeled." Therefore, it's not surprising that models have become crucial also in technical fields such as mechanics, civil engineering, and ultimately in computer science and computer engineering. Within production processes, modeling allows us to investigate, verify, document, and discuss properties of products before they are actually produced. In many cases, models are even used for directly automating the production of goods.

The discussion about whether modeling is good or bad is not really appropriate. We all always create a mental model of reality. In this sense, one can say that you cannot avoid modeling. This is even more appropriate when dealing with objects or systems that need to be developed: in this case, the developer must have in mind a model for her/his objective. The model always exists, the only option designers have is about its form: it may be mental (existing only in the designers' heads) or explicit [53]. In other words, the designer can decide whether to dedicate effort to realizing an explicit representation of the model or keeping it within her/his own mind.

1.2   MODELING FOR SOFTWARE DEVELOPMENT

The scope of this book is to discuss approaches based on modeling for the development of software artifacts. In this context, this is known as *Model-Driven Software Engineering* (MDSE). MDSE practices proved to increase efficiency and effectiveness in software development, as demonstrated by various quantitative and qualitative studies [1].

The need for relying on models for software development is based on four main facts.

1. Software artifacts are becoming more and more complex and therefore they need to be discussed at different abstraction levels depending on the profile of the involved stakeholders, phase of the development process, and objectives of the work.

2. Software is more and more pervasive in people's lives, and the expectation is that the need for new pieces of software or the evolution of existing ones will be continuously increasing.

3. The job market experiences a continuous shortage of software development skills with respect to job requests.

4. Software development is not a self-standing activity: it often imposes interactions with non-developers (e.g., customers, managers, business stakeholders, etc.) which need some mediation in the description of the technical aspects of development.

Modeling is a handy tool for addressing all these needs. That's why we strongly believe that MDSE techniques will see more and more adoption. This vision is supported by all the major players in this field (i.e., tool vendors, researchers, and enterprise software developers), and also by business analysts. For instance, Gartner foresees a broad adoption of model-driven techniques thanks to the convergence of software development and business analysis. This is particularly true in scenarios where the complexity of new, service-oriented architectures (SOAs) and cloud-based architectures, jointly applied with business process management (BPM), demand more abstract approaches than mere coding. Also from the standardization and tooling point of view, business and IT modeling and technologies are converging. This brings huge benefits to organizations, which struggle to bridge the gap between business requirements and IT implementation. Companies try to exploit this convergence and to ensure collaboration among the IT organization, business process architects, and analysts who are using and sharing models. Obviously, this also implies organizational changes that should move toward more agile approaches, combined with fostering modeling efforts and reusable design patterns and frameworks to improve productivity, while ensuring quality and performance [11].

A different discussion is the use people make of models (based on Martin Fowler's classi-fication[1]):

- models as sketches: models are used for communication purposes, only partial views of the system are specified;

- models as blueprints: models are used to provide a complete and detailed specification of the system; and

- models as programs: models, instead of code, are used to develop the system.

Of course, during the course of a development process, a team can use the models in several different ways. For instance, while discussing design decisions models could be used as sketches as an aid for the discussion; and after, complete models could be defined as part of the blueprint of the system. Finally, these blueprint models may be further refined to create the system using code generation techniques to minimize the coding tasks.

1.3   HOW TO READ THIS BOOK

This is an introductory book to model-driven practices and aims at satisfying the needs of a diverse audience.

---

[1]http://martinfowler.com/bliki/UmlMode.html

Our aim is to provide you with an agile and flexible book to introduce you to the MDSE world, thus allowing you to understand the basic principles and techniques, and to choose the right set of instruments based on your needs. A word of caution: this is neither a programming manual nor a user manual of a specific technology, platform, or toolsuite. Through this book you will be able to delve into the engineering methodologies, select the techniques more fitting to your needs, and start using state of the art modeling solutions in your everyday software development activities.

We assume the reader is familiar with some basic knowledge of software engineering practices, processes, and notations, and we also expect some software programming capabilities.

The book is organized into two main parts.

- The first part (from Chapter 2 to Chapter 6) discusses the *foundations of MDSE* in terms of principles (Chapter 2), typical ways of applying it (Chapter 3), a standard instantiation of MDSE (Chapter 4), the practices on how to integrate MDSE in existing development processes (Chapter 5), and an overview of modeling languages (Chapter 6).

- The second part (from Chapter 7 to Chapter 10) deals with the *technical aspects of MDSE*, spanning from the basics on how to build a domain-specific modeling language (Chapter 7), to the description of Model-to-Model and Model-to-Text transformations (Chapters 8 and 9, respectively), and the tools that support the management of MDSE artifacts (Chapter 10).

We may classify the attitude of a reader of this book into three main categories: the curious, the user, and the developer—plus, the atypical one of the B.Sc., M.Sc, or Ph.D. student or who needs to address some academic course on MDSE. It's up to you to decide what are your expectations from this book. Obviously, you may want to read the book from the first to the last page anyway, but here we try to collect some suggestions on what not to miss, based on your profile.

MDSE Curious

If you are interested in general in MDSE practices and you would like to learn more about them, without necessarily expecting to do the technical work of using model-driven approaches or developing modeling languages, you are probably *MDSE curious*.

This attitude is extremely useful for CTOs, CIOs, enterprise architects, business and team managers who want to have a bird's eye view on the matter, so as to make the appropriate decisions when it comes to choosing the best development techniques for their company or team.

In this case, our suggestion is to focus on the first part of the book, namely the: MDSE principles (Chapter 2) and usage (Chapter 3); MDA proposal by OMG (Chapter 4); and overview on the modeling languages (Chapter 6). We also suggest reading how to integrate MDSE in existing development processes (Chapter 5) and possibly the tools that support the management of MDSE artifacts (Chapter 10).

MDSE User

If you are a technical person (e.g., software analyst, developer, or designer) and you expect to use in your development activities some MDSE approach or modeling notation, with the purpose of improving your productivity and learning a more conceptual way for dealing with the software development problem, you are propably an *MDSE user*. In this case, you are most likely not going to develop new modeling languages or methods, but you will aim at using the existing ones at your best.

If you think you fall into this category, we recommend that you read at least the basics on MDSE principles and usage in Chapters 2 and 3; overview on OMG's MDA approach (Chapter 4) and on modeling languages (Chapter 6); description of Model-to-Model and Model-to-Text transformations (Chapters 8 and 9, respectively); and tools that support the management of MDSE projects (Chapter 10).

MDSE Developer

Finally, if you already have some basic knowledge of MDSE but you want to move forward and become a hardcore MDSE adopter, by delving into the problems related to defining new DSLs, applying end-to-end MDSE practices in your software factory, and so on, then we classify your profile as *MDSE developer*.

As such, you are probably interested mostly in the second part of the book, which describes all the details on how to technically apply MDSE. Therefore, we recommend that you read at least the following chapters: the overview on modeling languages (Chapter 6), the basis on domain-specific modeling languages (Chapter 7); the description of Model-to-Model and Model-to-Text transformations (Chapters 8 and 9, respectively), and the tools that support the management of MDSE artifacts (Chapter 10).

Optionally, if you need to refresh your mind on the basic MDSE principles, you can read Chapters 2 and 3. As a final read, if you are also involved in management activities, you can read the chapter on agile development processes (Chapter 5).

Student

If you are a *student* of an academic or professional course, our suggestion is to go through the whole book so as to get at least the flavor of the objectives and implications of the MDSE approach. In this case, this book possibly will not be enough to cover in detail all the topics of the course you are following. Throughout the text we offer a good set of references to books, articles, and online resources that will help you investigate the topics you need to study in detail.

Instructor

If you are an *instructor* of an academic or professional course, you will be able to target your specific teaching needs by selecting some chapters of the book depending on the time span available in your course and the level of technical depth you want to provide your students with. The teaching materials covering the whole book are available as slidesets, linked from the book web site (`http://www.mdse-book.com`).

In addition to the contents of the book, more resources are provided on the book's website (`http://www.mdse-book.com`), including the examples presented in the book.

CHAPTER 2

MDSE Principles

Models are paramount for understanding and sharing knowledge about complex software. MDSE is conceived as a tool for making this assumption a concrete way of working and thinking, by transforming models into first-class citizens in software engineering. Obviously, the purpose of models can span from communication between people to executability of the designed software: the way in which models are defined and managed will be based on the actual needs that they will address. Due to the various possible needs that MDSE addresses, its role becomes that of defining sound engineering approaches to the definition of models, transformations, and their combinations within a software development process.

This chapter introduces the basic concepts of MDSE and discusses its adoption and perspectives.

2.1 MDSE BASICS

MDSE can be defined as a methodology[1] for applying the advantages of modeling to software engineering activities. Generally speaking, a methodology comprises the following aspects.

- *Concepts*: The components that build up the methodology, spanning from language artifacts to actors, and so on.

- *Notations*: The way in which concepts are represented, i.e., the languages used in the methodology.

- *Process and rules*: The activities that lead to the production of the final product, the rules for their coordination and control, and the assertions on desired properties (correctness, consistency, etc.) of the products or of the process.

- *Tools*: Applications that ease the execution of activities or their coordination by covering the production process and supporting the developer in using the notations.

All of these aspects are extremely important in MDSE and will be addressed in this book.

In the context of MDSE, the *core concepts* are: models and transformations (i.e., manipulation operations on models). Let's see how all these concepts relate together, revisiting the famous equation from Niklaus Wirth:

---

[1]We recognized that methodology is an overloaded term, which may have several interpretations. In this context, we are not using the term to refer to a formal development process, but instead as a set of instruments and guidelines, as defined in the text above.

Algorithms + Data Structures = Programs

In our new MDSE context, the simplest form of this equation would read as follows:

Models + Transformations = Software

Obviously, both models and transformations need to be expressed in some *notation*, which in MDSE we call a modeling language (in the same way that in the Wirth equation, algorithms and data structures need to be defined in some programming language). Nevertheless, this is not yet enough. The equation does not tell us what kinds of models (and in which order, at what abstraction level, etc.) need to be defined depending on the kind of software we want to produce. That's where the model-driven *process* of choice comes to play. Finally, we need an appropriate set of *tools* to make MDSE feasible in practice. As for programming, we need IDEs that let us define the models and transformations as well as compilers or interpreters to execute them and produce the final software artifacts.

Given this, it goes without saying that MDSE takes the statement "everything is a model" very seriously. In fact, in this context one can immediately think of all the ingredients described above as something that can be modeled. In particular, one can see transformations as particular models of operations upon models. The definition of a modeling language itself can be seen as a model: MDSE refers to this procedure as *metamodeling* (i.e., modeling a model, or better: modeling a modeling language; or, modeling all the possible models that can be represented with the language). And this can be recursive: modeling a metamodel, or describing all metamodels, means to define a meta-metamodel. In the same way one can define as models also the processes, development tools, and resulting programs.

MDSE tries to make the point that the statement "everything is a model" has a strong unifying and driving role in helping adoption and coherence of model-driven techniques, in the same way as the basic principle "everything is an object" was helpful in driving the technology in the direction of simplicity, generality, and power of integration for object-oriented languages and technologies in the 1980s [10].

Before delving into the details of the terminology, one crucial point that needs to be understood is that MDSE addresses design of software with a *modeling* approach, as opposed to a *drawing* one. In practical terms, we distinguish these two approaches because drawing is just about creating nice pictures, possibly conforming to some syntactical rules, for describing some aspects of the design. On the other side, modeling is a much more complex activity that possibly implies graphical design (but that could be replaced by some textual notations), but it's not limited to depicting generic ideas: in modeling the drawings (or textual descriptions) have implicit but unequivocally defined semantics which allow for precise information exchange and many additional usages. Modeling, as opposed to simply drawing, grants a huge set of additional advantages, including: syntactical validation, model checking, model simulation, model transformations, model execution (either through code generation or model interpretation), and model debugging.

2.2   LOST IN ACRONYMS: THE MD* JUNGLE

The first challenge that a practitioner faces when addressing the model-driven universe is to cope with the plethora of different acronyms which could appear as obscure and basic synonyms. This section is a short guide on how to get out of the acronym jungle without too much burden. Figure 2.1 shows a visual overview of the relations between the acronyms describing the modeling approaches.

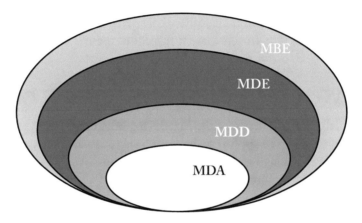

Figure 2.1: Relationship between the different MD* acronyms.

Model-Driven Development (MDD) is a development paradigm that uses models as the primary artifact of the development process. Usually, in MDD the implementation is (semi)automatically generated from the models.

Model-Driven Architecture (MDA) is the particular vision of MDD proposed by the Object Management Group (OMG) and thus relies on the use of OMG standards. Therefore, MDA can be regarded as a subset of MDD, where the modeling and transformation languages are standardized by OMG.

On the other hand, MDE would be a superset of MDD because, as the E in MDE suggests, MDE goes beyond the pure development activities and encompasses other model-based tasks of a complete software engineering process (e.g., the model-based evolution of the system or the model-driven reverse engineering of a legacy system).

Finally, we use "model-based engineering" (or "model-based development") to refer to a softer version of MDE. That is, the MBE process is a process in which software models play an important role although they are not necessarily the key artifacts of the development (i.e., they do NOT "drive" the process as in MDE). An example would be a development process where, in the analysis phase, designers specify the domain models of the system but subsequently these models are directly handed out to the programmers as blueprints to manually write the code (no automatic code generation involved and no explicit definition of any platform-specific model). In this process, models still play an important role but are not the central artifacts of the development process and may be less complete (i.e., they can be used more as blueprints or sketches of the system) than those in an MDD approach. MBE is a superset of MDE. All model-driven processes are model-based, but not the other way round.

All the variants of "model-driven whatever" are often referred to with the acronym *MD* (Model-Driven star)*. Notice that a huge set of variants of all these acronyms can be found in literature too. For instance, MDSE (Model-Driven Software Engineering), MDPE (Model-Driven Product Engineering), and many others exist. MDE can be seen as the superset of all these variants, as any MD*E approaches could fall under the MDE umbrella. The focus of this book is on MDSE.

2.3 OVERVIEW OF THE MDSE METHODOLOGY

In this section we delve into the details of the MDSE ingredients and philosophy. In particular, we will clarify that modeling can be applied at different levels of abstractions, and that a full-fledged modeling approach even leads to modeling the models themselves. Finally, we will describe the role and nature of model transformations.

2.3.1 OVERALL VISION

MDSE provides a comprehensive vision for system development. Figure 2.2 shows an overview of the main aspects considered in MDSE and summarizes how the different issues are addressed. MDSE seeks for solutions according to orthogonal dimensions: conceptualization (columns in the figure) and implementation (rows in the figure).

The *implementation* issue deals with the mapping of the models to some existing or future running systems. Therefore, it consists of defining three core aspects.

- The modeling level: where the models are defined.

- The realization level: where the solutions are implemented through artifacts that are actually in use within the running systems (this consists of code in case of software).

- The automation level: where the mappings from the modeling to the realization levels are put in place.

The *conceptualization* issue is oriented to defining conceptual models for describing reality. This can be applied at three main levels.

- The application level: where models of the applications are defined, transformation rules are performed, and actual running components are generated.

- The application domain level: where the definition of the modeling language, transformations, and implementation platforms for a specific domain are defined.

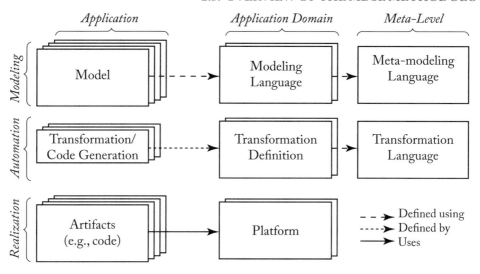

Figure 2.2: Overview of the MDSE methodology (top-down process).

- The meta-level: where conceptualization of models and of transformations are defined.

The core flow of MDSE is from the application models down to the running realization, through subsequent model transformations. This allows reuse of models and execution of systems on different platforms. Indeed, at the realization level the running software relies on a specific platform (defined for a specific application domain) for its execution.

To enable this, the models are specified according to a modeling language, in turn defined according to a metamodeling language. The transformation executions are defined based on a set of transformation rules, defined using a specific transformation language.

In this picture, the system *construction* is enabled by a *top-down process* from *prescriptive models* that define how the scope is limited and the target should be implemented. On the other side, *abstraction* is used *bottom-up* for producing *descriptive models* of the systems.

The next subsections discuss some details on modeling abstraction levels, conceptualization, and model transformations.

2.3.2   DOMAINS, PLATFORMS, AND TECHNICAL SPACES

The nature of MDE implies that there is some context to model and some target for the models to be transformed into. Furthermore, the software engineering practices recommend to distinguish between the problem and the solution spaces: the *problem space* is addressed by the *analysis* phase in the development process, while the *solution space* is addressed by the *requirements collection* phase first (defining *what* is the expected outcome), and subsequently by the *design* phase (specifying *how* to reach the objective). The common terminology for defining such aspects in MDSE is as follows.

The *problem space* (also known as problem domain) is defined as the field or area of expertise that needs to be examined to understand and define a problem. The *domain model* is the conceptual model of the problem domain, which describes the various entities, their attributes, roles, and relationships, plus the constraints and interactions that describe and grant the integrity of the model elements comprising that problem domain. The purpose of domain models is to define a common understanding of a field of interest, through the definition of its vocabulary and key concepts. One crucial recommendation is that the domain model must not be defined with a look-ahead attitude toward design or implementation concerns, while it should only describe assets and issues in the problem space.

On the contrary, the *solution space* is the set of choices at the design, implementation, and execution level performed to obtain a software application that solves the stated problem within the problem domain.

Technical spaces represent working contexts for the design, implementation, and execution of such software applications. These working contexts typically imply a binding to specific implementation technologies (which can be combined together into a coherent *platform*) and languages.

The concept of technical space is crucial for MDSE because it enables the possibility of deciding the set of technical tools and storage formats for models, transformations, and implementations. Notice that a technical space can either span both the problem and solution domains or cover only one of these aspects.

Figure 2.3 shows some examples of technical spaces, which span different phases of the development: MDSE, spanning from the problem definition down to the design, implementation, and even execution of the application (e.g., through model interpretation techniques); XML and the Java framework, which are more oriented toward implementation.

During the software development process it is possible to move from one technical space to another (as represented by the arrows in the figure). This implies the availability of appropriate software artifacts (called *extractors*) that are able to extract knowledge from a technical space and of others (called *injectors*) that are able to inject such knowledge in another technical space.

Notice also that the way in which models are transformed and/or moved from one technical space to another depends on the business objective of the development activity: indeed, MDSE can be applied to a wide set of scenarios, spanning from software development automation, to system interoperability, reverse engineering, and system maintenance. These aspects will be addressed extensively in Chapter 3.

2.3.3   MODELING LANGUAGES

As we will see in detail later, modeling languages are one of the main ingredients of MDSE. A modeling language is a tool that lets designers specify the models for their systems, in terms of graphical or textual representations. In any case, languages are formally defined and ask the

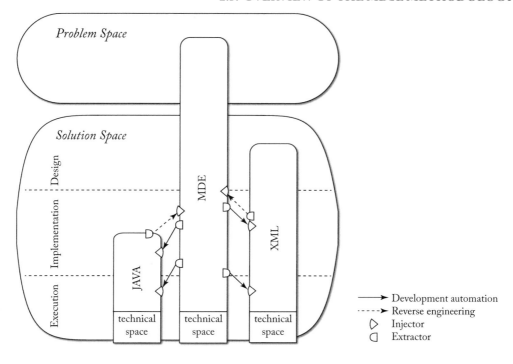

Figure 2.3: Technical spaces examples and coverage.

designers to comply with their syntax when modeling. Two big classes of languages can be identified.

Domain-Specific Languages (DSLs) are languages that are designed specifically for a certain domain, context, or company to ease the task of people that need to describe things in that domain. If the language is aimed at modeling, it may be also referred to as *Domain-Specific Modeling Language (DSML)*. DSLs have been largely used in computer science even before the acronym existed: examples of domain-specific languages include the well-known HTML markup language for Web page development, Logo for pen-based simple drawing for children, VHDL for hardware description languages, Mathematica and MatLab for mathematics, SQL for database access, and so on.

General-Purpose Modeling Languages (*GPMLs, GMLs,* or *GPLs*) instead represent tools that can be applied to any sector or domain for modeling purposes. The typical example for these kinds of languages is the UML language suite, or languages like Petri-nets or state machines.

To avoid misunderstandings and different variants on the naming, we will use DSL and GPL as acronyms for these two classes in the rest of this book.

Within these classes, further distinctions and classifications can be defined. Given that modeling inherently implies abstraction, a very simple way of classifying the modeling languages and the respective models is based on the *level of abstraction* at which the modeling is performed. Intuitively, it's easy to understand that some models are more abstract than others. In particular, when dealing with information systems design, one can think of alternative models that:

- describe requirements and needs at a very abstract level, without any reference to implementation aspects (e.g., description of user requirements or business objectives);

- define the behavior of the systems in terms of stored data and performed algorithms, without any technical or technological details;

- define all the technological aspects in detail.

Given the different modeling levels, appropriate transformations can be defined for mapping a model specified at one level to a model specified at another level.

Some methods, such as MDA, provide a fixed set of modeling levels, which make it easier to discuss the kinds of models that a specification is dealing with. In particular, MDA defines its abstraction hierarchy as from the list above.

Models are meant to describe two main dimensions of a system: the static (or structural) part and the dynamic (or behavioral) part. Thus, we can define the following:

- *Static models*: Focus on the static aspects of the system in terms of managed data and of structural shape and architecture of the system.

- *Dynamic models*: Emphasize the dynamic behavior of the system by showing the execution sequence of actions and algorithms, the collaborations among system components, and the changes to the internal state of components and applications.

This separation highlights the importance of having different views on the same system: indeed, a comprehensive view on the system should consider both static and dynamic aspects, preferably addressed separately but with the appropriate interconnections.

Without a doubt, multi-viewpoint modeling is one of the crucial principles of MDSE. Since modeling notations are focused on detailing one specific perspective, typically applying an MDSE approach to a problem may lead to building various models describing the same solution. Each model is focused on a different perspective and may use a different notation. The final purpose is to provide a comprehensive description of the system, while keeping the different concerns separated. That's why models can be interconnected through cross-referencing the respective artifacts.

While in principle, it's possible to define a design as composed of models based on several independent notations (possibly coming from different standardization bodies, or even including proprietary or third-party notations), it is convenient to exploit a suite of notations that, despite being different and addressing orthogonal aspects of the design, have a common foundation and are aware of each other. That's the reason why general-purpose languages typically do not include just one single notation, but instead include a number of coordinated notations that complement each other. These languages are also known as *Modeling Language Suites*, or *Family of Languages*, as they are actually composed of several languages, not just one. The most known example of language suites is UML itself, which allows the designers to represent several different diagram types (class diagram, activity diagram, sequence diagram, and so on).

2.3.4   METAMODELING

As models play a pervasive role in MDSE, a natural subsequent step to the definition of models is to represent the models themselves as "instances" of some more abstract models. Hence, exactly in the same way we define a model as an abstraction of phenomena in the real world, we can define a *metamodel* as yet another abstraction, highlighting properties of the model itself. In a practical sense, metamodels basically constitute the definition of a modeling language, since they provide a way of describing the whole class of models that can be represented by that language.

Therefore, one can define models of the reality, and then models that describe models (called metamodels) and recursively models that describe metamodels (called meta-metamodels). While in theory one could define infinite levels of metamodeling, it has been shown, in practice, that meta-metamodels can be defined based on themselves, and therefore it usually does not make sense to go beyond this level of abstraction. At any level where we consider the metamodeling practice, we say that a model *conforms* to its metamodel in the way that a computer program conforms to the grammar of the programming language in which it is written. More specifically, we say that a model conforms to its metamodel when all its elements can be expressed as instances of the corresponding metamodel (meta)classes as seen in Figure 2.4.

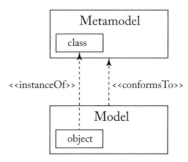

Figure 2.4: *conformsTo* and *instanceOf* relationships.

Figure 2.5 shows an example of metamodeling at work: real-world objects are shown at level M0 (in this example, a movie); their modeled representation is shown at level M1, where the model describes the concept of Video with its attributes (title in the example). The metamodel of this model is shown at level M2 and describes the concepts used at M1 for defining the model, namely: Class, Attribute, and Instance. Finally, level M3 features the meta-metamodel that defines the concepts used at M2: this set collapses in the sole Class concept in the example.

It's clear that there is no need for further metamodeling levels beyond M3, as they would always include only the Class concept.

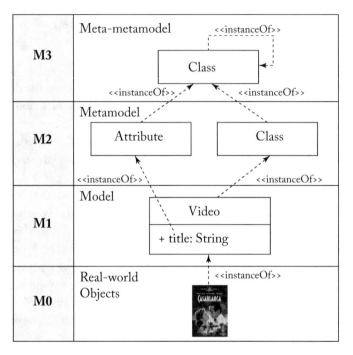

Figure 2.5: Models, metamodels, and meta-metamodels.

Although, one could think that metamodels represent a baroque conceptualization, meta-modeling is extremely useful in practice (even when it is not recognized as such). Metamodels can be used proficiently for:

- defining new languages for modeling or programming;
- defining new modeling languages for exchanging and storing information; and
- defining new properties or features to be associated with existing information (metadata).

Notably, the term *metadata* is defined with the same approach and aims at describing data about existing data. Therefore, metamodeling perfectly fits with the description of such information.

Notice that the MDSE paradigm, also known as *modelware*, is not so different from grammarware (i.e., the technical space where languages are defined in terms of grammars) in terms of basic definition and infrastructure. This is highlighted in Figure 2.6: grammarware (shown on the right-hand side of the picture) can be seen as a broad technical space for defining languages through a domain-specific syntax. One of the most common ways for defining a language is the

EBNF form, which is a textual representation of the grammar rules for the language (an alternative graphical representation also exists, known as syntax graph). A grammar definition language can be recursively defined based on itself (as represented in the cyclic arrow at the topmost level in the grammarware stack). A specific language conforms to its specification and a model (or program or text) can be defined according to the language. Analogously, in the MDSE stack a language definition facility is available, recursively defined based on itself, and all the language definitions can be specified according to it. Finally, all the models will conform to the specified language. The image describes this stack within the standardized MDA framework, as specified by OMG (we will explore this case in detail in Chapter 4).

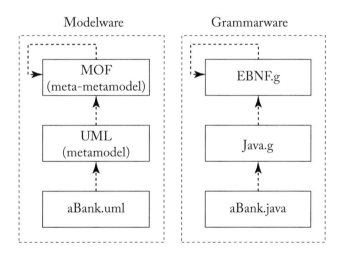

Figure 2.6: Modelware vs. Grammarware.

2.3.5   TRANSFORMATIONS

Besides models, model transformations represent the other crucial ingredient of MDSE and allow the definition of mappings between different models. Transformations are actually defined at the metamodel level, and then applied at the model level, upon models that conform to those metamodels. The transformation is performed between a source and a target model, but it is actually defined upon the respective metamodels (see also Figure 2.7).

MDSE provides appropriate languages for defining model transformations in order to provide designers with optimized solutions for specifying transformation rules. These languages can be used for defining model transformations in terms of transformation templates that are typically applied upon models according to some matching rules checked upon model elements.

Such transformation rules can be defined following different approaches: the transformation can be written manually from scratch by a developer, or can be defined as a refined specifica-

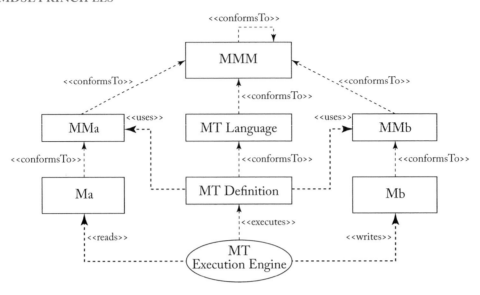

Figure 2.7: Role and definition of transformations between models.

tion of an existing one. Alternatively, transformations themselves can be produced automatically out of some higher level mapping rules between models. This technique is based on two phases:

1. defining a mapping between elements of a model to elements of another one (*model mapping* or *model weaving*); and

2. automating the generation of the actual transformation rules through a system that receives as input the two model definitions and the mapping between them and produces the transformations.

This allows the developers to concentrate on the conceptual aspects of the relations between models and then delegate the production of the transformation rules (possibly implemented within different technical spaces).

Another interesting aspect related to the vision of "everything is a model" is the fact that *transformations themselves can be seen as models*, and managed as such, including their metamodeling. The lower part of the Figure 2.7 shows two models (M_a and M_b), and a transformation M_t that transforms M_a into M_b. In the level above, the corresponding metamodels are defined (MM_a, MM_b, and MM_t), to which the three models (M_a, M_b, and M_t) conform, respectively. In turn, they all conform to the same meta-metamodel.

2.4 TOOL SUPPORT

Besides the conceptual tools provided by MDSE in terms of modeling and transformation languages, developers are provided with a large and diverse set of modeling tools that support the development of models, their transformation, and their integration within the software development process. Developers can nowadays choose among a complete spectrum of tools for MDSE, from open source to free to commercial ones, from full frameworks to individual tools, from desktop ones to cloud-based and SaaS (Software-as-a-service) solutions. Some of them support the whole MDSE process, in the sense that they provide facilities for defining new languages, models, transformations, and implementation platforms. Some others are design tools that support domain-specific notations or general-purpose ones.

2.4.1 DRAWING TOOLS VS. MODELING TOOLS

Many people assume that drawing tools and modeling tools are two interchangeable concepts but this is far from true. In fact, only some tools are drawing and modeling tools at the same time.

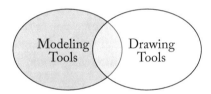

Figure 2.8: Modeling tools vs. drawing tools.

Some modeling tools use a concrete *textual syntax* for specifying models (not all models need to be graphical models), and therefore, there is no support for drawing them, even if the tools may be able to render the textual definition into some kind of graphical export format for visualization purposes. For instance, one can think about the textual modeling tools for UML[2] or any of the DSLs you may create with tools like XText[3] and EMFText.[4]

Furthermore, many drawing tools are not appropriate modeling tools. A drawing tool can be only considered as a modeling tool only if the tool "understands" the drawings, i.e., the tool does not deal with just shapes, lines, and arrows, but understands that they represent, e.g., classes, associations, or other modeling concepts. This should at least be enough to validate a model, i.e., check that the model is a correct instance of its metamodel.

For instance, a drawing tool may allow the designer to draw something funny like the model in Figure 2.9. The icons and shapes may be the ones defined in a given language (in the example, the UML notation) but the model may have no meaning at all, because the way elements are used

---

[2]http://modeling-languages.com/uml-tools/#textual
[3]http://www.eclipse.org/Xtext
[4]http://www.emftext.org

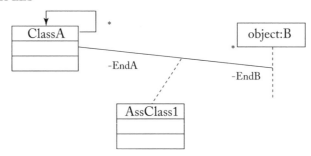

Figure 2.9: Wrong model defined with a drawing tool.

breaks the language rules. MDSE recommends always using a modeling tool to define models because of at least three main reasons.

- A modeling tool is able to export or manipulate the models using (external) access APIs provided by the tools. The API of a drawing tool may offer methods like *getAllRectangular-Shapes* but not a method like *getAllClasses*. The latter *is* the kind of API method exposed in a modeling tool, which is needed to easily manipulate models. The same discussion applies to the export formats. Drawing tools hardly offer any kind of semantic-aware export format that can be easily imported into other (modeling) tools.

- A modeling tool guarantees a minimum level of semantic meaning and model quality, because it grants alignment to some kind of metamodel. For the same drawing effort, having a model as an outcome (and not just a drawing) is much more valuable, even if the model has been defined just for communication purposes. In fact, it is very likely that this model could become useful for prototyping or code generation too, and for those purposes you must be able to use the model as a part of an MDSE chain.

- A modeling tool typically provides appropriate features for model transformations. Obviously, since at the end models are encoded as files, one can think to use usual imperative programming languages for defining model transformations. However, this lowers the level of abstraction of the entire modeling framework and typically ends up in producing cumbersome and unmaintainable pieces of software. That's why MDSE pushes for adopting declarative model transformation languages that feature a set of primitives explicitly targeted to defining the transformation rules.

2.4.2 MODEL-BASED VS. PROGRAMMING-BASED MDSE TOOLS

One initial distinction that can be made between MDSE tools is their classification in terms of the development paradigm used for their implementation. Indeed, MDSE tools themselves can be developed using traditional coding techniques or by applying the same MDSE principles that they aim to promote and support. Although both approaches are valid, in this book we will put more emphasis on model-based MDSE tools, due to their better integration, conceptual simplicity, and also intellectual coherency and honesty (in a sense, one should apply the principles one advocates).

2.4.3   ECLIPSE AND EMF

Besides the various proprietary modeling tools that are available for model-based enterprise development, one tooling platform that has become prominent in the MDSE world is the Eclipse development environment. A set of interesting tools for MDSE have been made available under Eclipse, thus enabling a fertile flourishing of initiatives upon this platform. In this book we will mainly refer to the Eclipse framework since it's open source (but allows commercial extensions) and comprises popular components for all the modeling tasks we will describe.

The Eclipse Modeling Framework (EMF) is the core technology in Eclipse for model-driven engineering. EMF is a good representative of model-based MDSE tools for various reasons. First, EMF allows the definition of metamodels based on the metamodeling language called *Ecore*. Second, EMF provides generator components for producing from metamodels (i) a specific Java-based API for manipulating models programmatically and (ii) modeling editors to build models in tree-based editors. Third, EMF comes with a powerful API covering different aspects such as serializing and deserializing models to/from XMI as well as powerful reflection techniques. Fourth, based on EMF, several other projects are defined which provide further functionalities for building model-based development support within Eclipse.

2.5   ADOPTION AND CRITICISMS OF MDSE

Modeling is frequently prescribed and advocated by software engineering practitioners and consultants. However, its adoption in industry is less frequently undertaken [6]. Some studies done in the past years show that modeling is not extensively adopted among practitioners [20] or it is not a core part of the development process [24], although more recent studies have a more positive view [34].

It is true that modeling has yet to cross the chasm and become mainstream, mainly due to inflated expectations raised in the past by various MDSE initiatives, including MDA and before that plain UML-based proposals. However, new developments in the last few years in terms of new technologies and methods (like domain-specific languages), the widening of the scope of the field (modeling is not only used for code generation but also for reverse engineering, interoperability, and software maintenance), a more pragmatic modeling approach (no need to model every detail for every project), and the release of *de facto* standard tools (in particular the EMF open-source modeling framework on top of the Eclipse platform) have changed the landscape and suggest that mainstream industrial adoption may happen sooner than later. As Steve Mellor likes to say, "modeling and model-driven engineering will be commonplace in three years time."[5]

---

[5]With the caveat that, as he said, he has been giving the same answer since 1985!

As with basically any other new technology/paradigm, MDSE is following the typical technology hype cycle (see Figure 2.10), where after an initial peak of inflated expectations (mostly caused by the popularization of UML, sold by some vendors and consultancy companies as the silver bullet for software development), MDSE fell down to the trough of disillusionment. We are now at the slope of enlightenment where our better knowledge of what MDSE is for and how it must be used helps companies to choose better the projects where MDSE can make a difference so that we can soon achieve the plateau of productivity where companies will be really able to take profit from the benefits MDSE brings to the table. We hope this book contributes to this evolution.

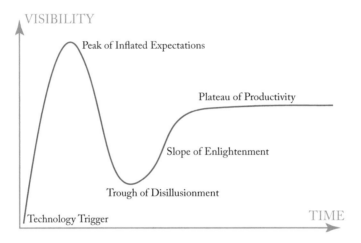

Figure 2.10: Technology hype cycle.

At this stage, adoption of MDSE still offers plenty of opportunities and a competitive advantage over more conservative organizations that will wait until the end before jumping on the bandwagon (Figure 2.11).

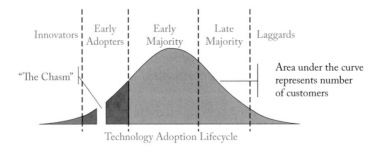

Figure 2.11: Technology adoption life cycle.

Despite the technical change that companies must face when embracing MDSE, experts often think that the biggest challenges faced by companies adopting model-driven approaches are not so much on the technical side, but much more on the human factor side [34]. Here are some responses by well-known experts in the field when asked about the biggest difficulty in embracing MDSE, BPM, SOA, and similar practices:

- "dealing with people's habits and resistance to change" (R. M. Soley, OMG);

- "accepting and adopting standards" (S. White, IBM);

- "who to ask for guidance and training, and tools availability; difficulty of communication between users, business lines, and other stakeholders" (A. Brown, The Open Group);

- "dealing and managing people" (S. Mellor); and

- "impatience of getting to the results" (T. Jensen, IBM).

These aspects sum up to the well-known fact that: "Learning a new tool or technique actually lowers programmer productivity and product quality initially. You achieve the eventual benefit only after overcoming this learning curve" [29].

Despite the advantages of MDSE, one should not regard it as the panacea for all the problems in the world. Some experts actually have some words of caution about the MDSE approach and what it implies:

- Beware of statements of pure principles: when it comes down to it, the real point of software engineering practice is to increase productivity, reduce errors, and cut code [25]. When MDSE is not perceived as delivering such properties, it is regarded with suspicion.

- Sometimes, the main message of MDSE is perceived as advocating modeling with respect to programming. However, that is not the right way to see it. The real question instead is to understand and define the right abstraction level for addressing each development activity. At the end, this applies also to programming itself.

- In the view of traditional programmers, diagrams are considered, after all, just pretty pictures. As such, they represent more of a burden than an advantage (as for documentation, they need to be kept aligned, discussed, and so on). This opinion is mainly due to unclear understanding of modeling (vs. drawing) and of its advantages, as we will discuss later in the book.

- Models may be perceived as an oversimplification of reality [63]. Indeed, models are an abstraction of reality, where unnecessary details are omitted. However, MDSE practices can be used to combine or refine models to cover all the interesting aspects.

- Models are perceived as useless artifacts that nobody will appreciate, including end users that just look for software that executes and performs well. This is a way to express that models and running applications are perceived as competitors, which is not true at all, especially if you consider the concept of executable models, which can become such through code generation or model-interpretation techniques.

- "To a Computer Scientist, everything looks like a language design problem. Languages and compilers are, in their opinion, the only way to drive an idea into practice" (David Parnas). This is a clear warning on how and when to apply MDSE techniques for language definition. As we will see later when addressing domain-specific languages, we also advise prudence in embarking in a language definition task.

CHAPTER 3

MDSE Use Cases

The first and most known application scenario for MDSE is definitely the one of *software development automation* (typically known as model-driven development (MDD)) where model-driven techniques are consistently employed with the goal of automating as much as possible the software lifecycle from the requirements down to the deployed application. However, the full spectrum of possible applications for MDSE spans a much wider set of scenarios; MDD is just the tip of the MDSE iceberg (as shown in Figure 3.1).

In this sense, the acronym MD(S)E can also be read as *Model-Driven Everything*; that is, the MDSE philosophy can be applied to any software engineering task. This chapter will focus on three main application scenarios that illustrate how the benefits of MDSE can apply to different needs. Apart from the software development scenario, we will describe the application of MDSE to the reverse engineering and software modernization field, to the tool/systems' interoperability problem and we will touch as well the use of models in an organization, beyond purely its software systems.

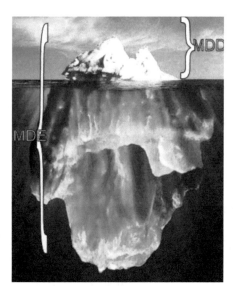

Figure 3.1: MDD is just the most visible side of MDSE.

3.1   AUTOMATING SOFTWARE DEVELOPMENT

Software development automation consists of starting from a high level (or early) representation of the desired software features and deriving a running application out of it, possibly through a set of intermediate steps to enable some degree of user interaction with the generation process.

Typically, when following an MDSE approach, the running application can be obtained through one or more model transformations that subsequently produce a more and more refined version of the software description, until an executable version of it is reached. A usual MDSE-based development process is illustrated in Figure 3.2. In each phase, models are (semi)automatically generated using model-to-model transformations taking as input the models obtained in the previous phase (and manually completed/refined when necessary). In the last step, the final code is generated by means of a model-to-text transformation from the design models.[1]

Figure 3.2: A typical MDSE-based software development process.

There are other benefits of introducing MDSE into the development process. One of the main advantages is to bridge the communication gap between requirements/analysis and implementation. At the organization level, this corresponds to bridging the gap between business needs and IT realization or support. This is a major problem in a number of enterprises, and models have proven to be a valid solution as a *lingua franca* among actors from business and IT divisions.

Furthermore, models capture and organize the understanding of the system in a way that facilitates the discussion among team members (and eases the integration of new ones), are well suited for documentation, permit earlier exploration of design alternatives and the early evaluation of the system adequacy, increase the decomposition and modularization of the system, and improve the reuse of parts of the system in new projects and the system evolution and maintenance (e.g., by facilitating tracing back the code to the original requirements). Overall, the most direct benefits of MDE can be summarized as the increase of communication effectiveness between the stakeholders and increase in the productivity of the development team thanks to the (partial) automation of the development process. As a side effect, this automation reduces also the number of defects in the final code that could be inadvertently introduced by the developers.

In order to be able to generate a running system from the models, they must be executable.[2] An *executable model* is a model complete enough to be executable. From a theoretical point of view, a model is executable when its operational semantics are fully specified. In practice, the executability of a model may depend more on the adopted execution engine than on the model itself. On the one hand, we may find some models which are not entirely specified but that can be executed

---

[1]As we will discuss later on, this is just an example. In some contexts we can go directly from the analysis models to the code or, when using model interpretation, skip the implementation phase and execute directly the models.

[2]This doesn't mean that all models created during the process have to be executable but at least some of them must.

by some advanced tools that are able to "fill the gaps" and execute them; on the other hand, we may have very complex and complete models that cannot be executed because an appropriate execution engine is missing. For instance (Figure 3.3), given a simple class diagram specifying only the static information of a domain, a trivial code generator will only be able to generate the skeletons of the corresponding (Java) classes while a more complex one would be able to infer most of the system behavior out of it. As an example, this advanced generator could assume that for each class in the diagram the system will need to offer all typical CRUD (create/read/update/delete) operations and thus, it could directly decide to create all the forms and pages implementing these operations.[3] In fact, studies like [3] show that the CRUD operations account for a staggering 80% of the overall software functionality in typical data-intensive applications so this advanced code generator could save a lot of development time from a minimal system specification.

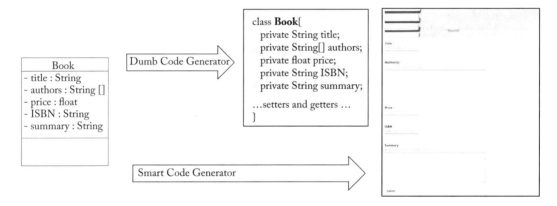

Figure 3.3: Code generation example.

The most well-known family of executable models are those based on the UML language, generically referred as *executable UML*.[4] Executable UML models make extensive use of an action language (kind of imperative pseudocode) to precisely define the behavior of all class methods, state transitions, etc. The OMG itself has recently standardized this notion of executable UML models. In particular, the OMG has standardized the semantics of a Foundational Subset for Executable UML Models (fUML) that suffices to model software specifications in a way that makes them suitable to be the input of an automated software development process. The adopted action language for fUML is known as the *Action Language for fUML*, or Alf.[5] Alf is basically a textual notation for UML behaviors that can be attached to a UML model at any place a fUML

---

[3]Note that this is the strategy that many MVC frameworks for popular languages like Ruby, PHP, or Python already follow; frameworks like Ruby on Rails, Symfony or Django automatically create a production-ready administration web interface based on the model's definition.

[4]Executable UML is also used to refer to the specific executable UML development method (xUML) proposed by S. Mellor [48] as an evolution of the previous Shlaer-Mellor method.

[5]See http://www.omg.org/spec/ALF for the Alf Language specification or http://modeling-languages.com/new-ex ecutable-uml-standards-fuml-and-alf for a short introduction by Ed Seidewitz, the main language designer.

behavior can be. Syntactically, Alf looks at first much like a typical C/C++/Java legacy language, which softens its learning curve.

Code generation and model interpretation are then two different alternative strategies to "implement" execution tools and thus make executable models actually execute.

3.1.1   CODE GENERATION

Code generation aims at generating running code from a higher level model in order to create a working application, very much like compilers are able to produce executable binary files from source code. In this sense, code generators are also sometimes referred to as *model compilers*.

This generation is usually done by means of a rule-based template engine, i.e., the code generator consists in a set of templates with placeholders that once applied (instantiated) on the elements in the model, produce the code.

Once the code is generated, common IDE tools can be used to refine the source code produced during the generation (if necessary), compile it, and finally deploy it. As depicted in Figure 3.4, the single goal of a code generation process is to produce a set of source files from a set of models. All other activities around these files are performed using the same tools you would use to manipulate manually written code.

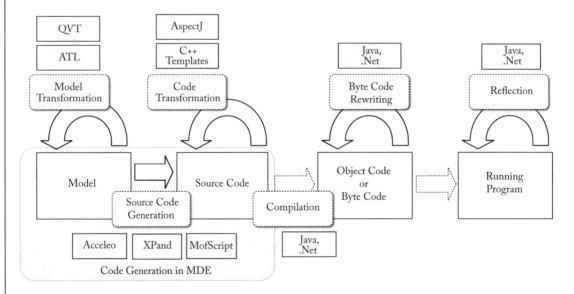

Figure 3.4: Scope of MDSE code generation (based on [71]).

Obviously, by "source code" we do not restrict ourselves to programming languages. By means of code generation, we can transform many kinds of models to many types of software artifacts (test cases, make files, documentation, configuration files, etc).

Partial vs. Full Generation

When the input models are not complete and the code generator is not smart enough to derive or guess the missing information, we can still benefit from a code generation approach by creating a partial implementation of the system. Nevertheless, partial code generation must be approached with caution. Partial generation means that programmers will need to complete the code manually to obtain a fully functional system. This leads to a situation where no single source of information exists: both the models and the code contain important information that may not be replicated in the other artifact, but at the same time some pieces of information appear in both places (e.g., the parts of the model that have been directly translated into the code). Having the same information in two places (code and models) is a recipe for trouble.

Strategies to maximize the benefit of partial code generation include:

- defining protected areas in the code, which are the ones to be manually edited by the developer. Thanks to this, a code generation tool can still regenerate the code with the assurance that the manually added code excerpts will be preserved;

- using round-trip engineering tool, so that changes on the code are, when possible, immediately reflected back to the model to keep both in sync; and

- focusing on complete generation of parts of the system more than on a partial generation of the full system. For instance, it is better to be able to say that the code generator creates the full database implementation of the system even if no generation is done on the business logic part than having a code generator that produces a little bit of both. We can regard the generation approach as full generation for some parts of the system avoiding the above risks for those parts.

Both, in full and partial generation scenarios, one objective to keep in mind is to try to generate the smallest possible amount of code to achieve the desired functionality. An effective code generation strategy should rely on all kinds of existing frameworks, APIs, components, or libraries in order to minimize the amount of actually generated code.

Advantages of Code Generation

There are several advantages that can be mentioned for the code generation approach.

- It protects the intellectual property of the modeler, as it allows the generation of the running application for a specific client, without sharing the conceptualization and design, which is the actual added value of the work and which can be reused or evolved in the future for other projects. Customers are starting to request the source code of the applications they pay for but so far there is no agreement regarding whether the models should be delivered as well.

- The generated implementation is model- and process-agnostic, and therefore easier to understand: the generated code is produced in a standard programming language that any developer can understand.

- Code generation allows customers to choose their runtime environment. This implies that customers can avoid any risk of vendor lock-in with respect to the MDSE technology that is used. Since standard code is produced, customers can even opt for abandoning the MDSE approach in the future.

- It can target a specific architecture based on the needs or legacy solution that the customer already has in house. The generated code can precisely follow the guidelines of the customer and can provide better integration with other IT components in the organization. In a sense, code generation grants two degrees of freedom instead of one: the model and the generator. Each of them can be tailored to the needs of the customer.

- Flexibility in the deployment of the generated code facilitates customers being compliant to their IT architecture policies and to other rules, also imposed by national or local laws (e.g., in terms of security, privacy, reliability, fault-tolerance, etc.) by exploiting the capabilities of a standard execution environment, without having to rely on a proprietary solution.

- Code generation allows reusing existing programming artifacts. In particular, existing pieces of code can be generalized and used as templates for code generation of the new parts of the software. If the code generator is flexible enough, it can be iteratively extended to generate more or richer pieces of code.

- A code generator is usually easier to maintain, debug, and track because it typically consists of rule-based transformations, while an interpreter has a generic and complex behavior to cover all the possible execution cases.

- Possibility of following a partial generation approach.

- In general, a generated application performs better in terms of execution speed with respect to a corresponding interpreted version.

Code generation is not free from a few drawbacks (c.f. the benefits of the model interpretation alternative as discussed in the next section). One of the big hurdles in adopting code generation techniques is the fact that the produced code does not look "familiar" to the developers. Even if it behaves exactly as expected and it is written in the programming language of choice, it could be very different from the code that actual programmers would write and this makes them reluctant to accept it (for instance, they may not feel confident to be able to tune it or evolve it if needed in the future). In this sense, it may be useful to define a *Turing test for code generation tools*. Similar to the classic Turing test for AI,[6] the Turing test for code generation tools reads as follows.

---

[6]http://en.wikipedia.org/wiki/Turing_test

A human judge examines the code generated by one programmer and one code generation tool for the same formal specification. If the judge cannot reliably tell the tool from the human, the tool is said to have passed the test.

Tools passing the test would have proved they can generate code comparable to that of humans and should be acceptable to them.

The chances of passing the Turing test for code generation may be increased by building the code generation templates starting from existing code manually written by the developers (generalizing and abstracting the templates from this code). Thus, when running these templates, they should produce code which looks more familiar to the developers.

3.1.2   MODEL INTERPRETATION

Model interpretation does not generate code from a model to create a working software application. Instead, a generic engine is implemented, which parses and executes the model on the fly, with an interpretation approach (exactly as interpreters do for interpreted programming languages).

Model interpretation is characterized by the following properties and advantages:

- It enables faster changes in the model because it does not require any explicit code generation step. This can lead to a significant shortening of the turnaround time in incremental development approaches, as the model can be run and modified on the fly.

- It even allows changing the model at runtime without stopping the running application, as the interpreter would continue the execution by parsing the new version of the model.[7]

- In principle, it enables the portability of the applications because it is possible to create an interpreter which is available on multiple platforms (e.g., multiple operating systems, cloud platforms, or technologies). The interpreter basically realizes an agnostic virtual environment where a modeled application can run, exactly in the same way an interpreter works on interpreted programming languages. To achieve the same with code generation, one should generate interpreted code, for which interpreters are available for different platforms.

- In case of model interpretation you don't need to (and you cannot) delve into the source code of the application anymore, simply because such a concept does not exist anymore (though some tools still allow you to add custom behavior as black-box components to be "interpreted" with the model depending on certain conditions). One could say that the model *is* the code, because the former actually replaces in all senses the latter.

- It enables the empowerment or modification of the behavior of the running application by updating the interpreter and still keeping the same models (the same can be achieved in case

---

[7]Models at runtime are also heavily employed in self-adaptive systems, i.e., systems that monitor the environment and are able to modify its behavior in response to changes on the environmental conditions. The information about how to adapt to these changes is usually expressed as a model that it is interpreted by the system at runtime.

of code generation by updating the generator, but then the code needs to be regenerated, compiled, and deployed again).

- It provides a higher level of abstraction (implemented by the interpreter) upon the system, according to the Platform-as-a-Service (PaaS) philosophy. Notice that the provided platform must comprise not only an execution engine able to run the model and produce the final state, but also many other tools (e.g., model debuggers, profilers, and so on) to fully exploit the potential of MDSE.

- It allows easy debugging of models at runtime, because interpretation can proceed step by step. This is natural for model interpretation, while it requires very sophisticated tools in case of code generation (because the executable application must be hooked to the model concepts, and the modeling tool must catch events from the running application).

- No deployment phase is needed, because the model is already the running version of the application.

Despite these benefits, this option still "scares" many MDSE users. The fact that the application source code is not available makes you dependent from the MDSE tool vendor. If it disappears you cannot just rely on the source code to continue executing and evolving the application. Performance concerns are also often cited to discard this option (both due to the fact that the model is interpreted and the limitations you have in tuning the application since you don't have full control on it), although this is not a real issue for most applications. Besides, with model interpretation customers are required to install a new proprietary platform in their IT infrastructure (the interpreter). This may be extremely critical in large organizations that have strict IT architecture policies and rely on separate platforms for managing non-functional requirements like security, reliability, modularity (e.g., think about information systems based on a service bus), etc.

3.1.3   COMBINING CODE GENERATION AND MODEL INTERPRETATION

Both code generation and model interpretation are used in practice, and not necessarily as mutually exclusive alternatives. Hybrid approaches are often used, either intertwined in the development process based on the experience of the developers or within a development platform as a combined solution.

For instance, within the MDSE process a developer may choose code generation to take an application to production, but at the same time the developer can adopt model interpretation during development time, e.g., to speed up the design of functional prototypes of the system thanks to the use of model simulations that help the developer to get a better understanding of the domain. Besides, a specific MDSE platform can opt for a hybrid code generation plus model interpretation approach. Some possibilities are:

- to provide a model interpretation approach based on an internal code generation strategy. This means that the tools actually generate, compile, and execute the code. However, this is hidden from the designer, who "feels" using a model interpretation approach (but with the option to partially configure the generation); and

- to provide a code generation oriented strategy that relies on predefined runtime components or frameworks to drastically reduce the amount of code to be generated. The predefined components might be flexible enough to carry out complex tasks, which need only simple pieces of code to be generated on purpose. This makes the code generation phase simpler and may produce a more readable source code (the source code can remain at a higher-abstraction level and delegate to the components most of the internal complexity). Ultimately, the "code" could simply be a set of XML descriptors and configuration files that could be "interpreted" by the runtime components. This dramatically reduces the complexity of building the interpreters with respect to the problem of building a generic interpretation for a complex modeling language. This solution provides the additional advantage of not requiring any compilation of the generated artifacts, which also remain easily readable by humans.

To summarize, there is not an approach better than the other in absolute terms. Choosing one approach or the other (or a combination of both) depends on the attitude and expertise of the work team, on the requirements of the application that is being built, and sometimes on the phase of the development process.[8]

3.2   SYSTEM INTEROPERABILITY

Interoperability is formally defined by IEEE as "the ability of two or more systems or components to exchange information and to use the information that has been exchanged" [28].

Interoperability is required in several scenarios: forward and reverse engineering (e.g., between two consecutive tools in the development chain), tool and language evolution (to address backward compatibility with previous versions), collaborative work (several members of the same organization may need to work together in the same task, even if they use two completely different systems to perform the task), system integration (e.g., when, after a business acquisition, the information system of the acquired company must communicate with the one of the parent company), and so forth.

Unfortunately, interoperability is a challenging problem that requires addressing both syntactic and semantic issues since each tool/component/system may use a different syntactic format to store its information and, more importantly, use its own internal interpreter (with its own semantics) to represent and manipulate such information, most likely different from the one expected by other tools. Therefore, trying a manual solution is error-prone and very time consuming,

---

[8]See also the discussion thread in: http://www.theenterprisearchitect.eu/archive/2010/06/28/model-driven-development-code-generation-or-model-interpretation.

and it is hardly reusable even when using a similar set of components. Instead of ad-hoc solutions, a generic set of bridges between the components of both systems should be provided. Each bridge should ensure data-level interoperability (i.e., metadata/data interchange) and operational-level interoperability (i.e., behavior interchange) between the parties, independently of the specific project/context in which they are used.

Initial attempts to build these bridges tried to implement bridges by directly connecting the components' APIs or interfaces. However, this low-level view of the systems was too limited to achieve real data interoperability. With the advent of MDSE, it has become possible to address the interoperability problem at a higher abstraction level.

Model-driven interoperability (MDI) approaches aim at defining bridges to achieve interoperability between two or more systems by applying model-driven techniques. They work by first making explicit the internal schema (i.e., metamodel) of each system (or each externally accessible component of the system). Metamodels are then aligned by matching the related concepts. Finally, model-to-model transformations exploit this matching information to export data (i.e., models) created with the first component to data conforming to the second component's internal schema. Deducing this internal schema can be performed following different strategies. For some systems the internal schema may be already available as part of the system specification, in the form of explicit models and associated documentation. For others, it can be derived from an analysis of the structure of the storage format used by the system (e.g., the XML schema in case the component stores XML data or a relational schema if the components manages information in a relational database management system). It is also possible to generate the metamodel from an analysis of the public API of the components [36].

A generic bridge to achieve interoperability between two systems A and B using MDSE as a pivot is depicted in Figure 3.5 (see [15] for a concrete application of this bridge). The basic principle of MDI is to decompose the bridge in two main parts: a syntactic mapping and a semantic mapping.

Syntactic Mapping

Syntactic mapping aims at crossing different technical spaces. The idea is to use *projectors* to go from a generally heterogeneous world (in terms of formats, techniques, etc.) to a more homogeneous world, in our case the modeling world of MDSE, and vice versa. Once in the modeling world, models can be used as the *lingua franca* between tools. These projectors can be directly hard coded using a general-purpose language such as Java, or preferably implemented with model-to-text or text-to-model transformation technologies (assuming that the input/output files are in text format), as those which we will see in the next chapters.

As mentioned in Chapter 2, there are two kinds of projectors:

- injectors that allow "injecting" the content of input files, using dedicated formats, as models (e.g., to transform the input file a_1 into model a_1 in Figure 3.5); and

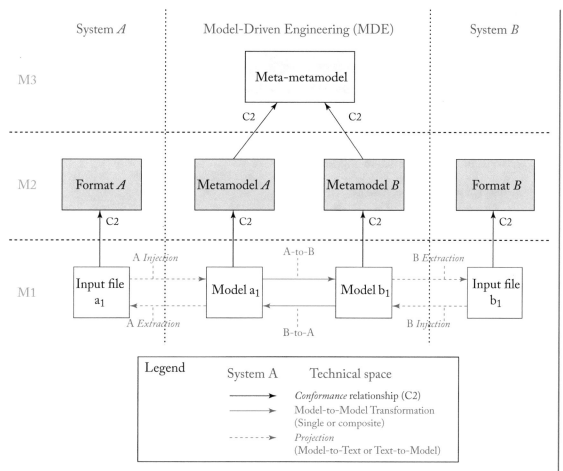

Figure 3.5: Generic interoperability bridge between two systems.

- extractors that allow "extracting" the content of models as output files using dedicated formats (e.g., to generate the output file b_1 from model b_1 in Figure 3.5).

Semantic Mapping

Semantic mapping aligns the concepts coming from the domains of both systems. This mapping is implemented as a model-to-model transformation. The transformation re-expresses the domain concepts of system A into a set of equivalent domain concepts understandable by system B. Metamodels A and B can be manually generated, derived from the corresponding format description (e.g., an XML schema when the input or output are XML documents), or automatically created if the formats to bridge conform to a meta-format for which a bridge at the metametalevel is already available. For instance, a generic bridge between the XML schema language and the

MDSE meta-metamodel would allow an automatic transformation of all format descriptions, i.e., specific XML schemas, to metamodels.

Overall View of MDI

To summarize, a standard MDI bridging process is actually composed of three main consecutive parts:

1. injection (text-to-model);

2. transformation (model-to-model); and

3. extraction (model-to-text).

All the actual domain knowledge is specified in the "transformation" part. The "projection" parts (i.e., "injection" or "extraction") only deal with technical/syntactical details, mainly concerning the storage formats.

Both projections and transformations can be either atomic or composite (i.e., chains of transformations). In most cases, such transformation chains are actually required in order to be able to split the overall problem into simpler ones, in order to provide better extensibility, reusability, and maintainability for the bridge.

3.3   REVERSE ENGINEERING

Nowadays, almost all organizations, independent of their size and type of activity, are facing the problem of managing, maintaining, or replacing their existing systems. These *legacy systems* are often large applications playing a critical role in the company's overall information system. They have been in use for a long time, they have been developed with now obsolete technology, and sometimes they are not completely documented. Therefore, the first problem to be solved when dealing with the evolution and/or modernization of legacy systems is to really understand what their architecture, provided functionalities, handled data, and enforced business rules and processes actually are.

This process of obtaining useful higher-level representations of legacy systems is commonly called *reverse engineering*, and remains today an open problem. MDSE comes to the rescue also in this scenario: Model-Driven Reverse Engineering (MDRE), i.e., the application of Model-Driven Engineering principles to provide relevant model-based views on legacy systems, has been recently proposed as a new solution to this important challenge.

The main goal of MDRE is to offer better support for the comprehension of existing systems thanks to the use of model-based representations of those systems. Taking as input the set of artifacts associated to the various components of the legacy system (spanning from source code, configuration files, databases, partial documentation, and so on), MDRE aims to create a set of models that represent the system (where each model focuses on a separate aspect of the system at a different abstraction level). These models can then be used for many different purposes, e.g., metrics and quality assurance computation, documentation generation, and tailored system view-points, or as a first step in a software evolution or modernization process, as shown in Figure 3.6.

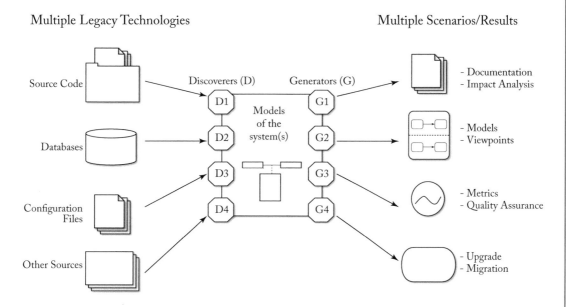

Figure 3.6: Model-driven reverse engineering.

An MDRE process includes three main phases.

- *Model Discovery.* In MDRE, the idea is to switch as soon as possible from the heterogeneous real world (with many legacy artifacts of a different nature) to the homogeneous world of models, where all artifacts are represented as a set of interrelated models. This is what we call the *model discovery* phase. A good approach for creating these models is to first focus on quickly creating a set of initial models that represent the legacy system at the same (low) abstraction level, i.e., without losing any information in the process. These *raw models* can be directly manipulated with the arsenal of model representation and manipulation techniques which we present in this book and, at the same time, are detailed enough to be the starting point of any of the MDRE applications outlined in Figure 3.6.

- *Model Understanding.* Most MDRE applications require the processing of the raw models discovered in the previous phase in order to obtain higher-level views of the legacy systems that facilitate their analysis, comprehension, and later regeneration. Thus, the second phase is the *model understanding* phase where chains of model manipulation techniques are employed to query and transform the raw models into more manageable representations (e.g., by omitting details that are not relevant for the MDRE scenario at hand or by deriving and making explicit the information hidden in the system structure, which helps to better understand the system itself).

- *Model (Re)Generation.* The processed models obtained at the end of the model understanding phase are finally used to generate and/or display the expected outcome of the reverse engineered process (e.g., the code of a refactored version of the system).

Figure 3.7 sketches a typical MDRE scenario where the aim is to migrate and redeploy an existing COBOL system into a Java platform.

In the first step (model discovery), we obtain the low-level models containing the full details of the COBOL code.

In the next step (model understanding), we raise the abstraction level of those models using model-to-model transformations and generating, for instance, UML diagrams with the domain information and rules extracted from the COBOL models. These models are then refactored or changed in various ways (e.g., to adapt them to the new policies in place in the organization or to adapt to new non functional requirements such as security or performance). The refactorings can be also implemented as model-to-model transformations when having UML models as both input and output of the transformation.

Subsequently, the models are used to generate the new running Java application. This is obtained through the third step (model generation), which transforms back the models to a lower-level representation if required (in this case, Java models) through an M2M transformation and then produces the proper Java code through a model-to-text transformation.

The Eclipse-MDT Modisco project[9] offers several model discoverers, generators, and transformations that can be used as building blocks for your reverse engineering projects.

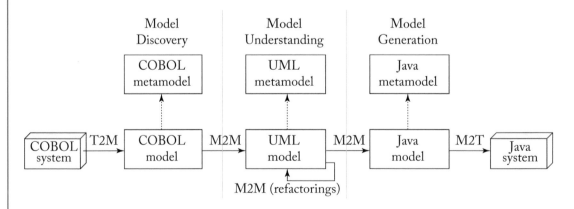

Figure 3.7: Example of a software modernization scenario.

---

[9]http://www.eclipse.org/MoDisco

3.4   MODELING THE ORGANIZATION

Previous scenarios have shown the benefits of MDSE when dealing with the software systems in your organization, but, in fact, models are useful in any other area of your organization. You can model your organization's goals, its internal structure, the relationships you have with your providers and clients …everything. And there are specific languages (what we call domain-specific languages, cf. Chapter 7) to model each one of these aspects.

In particular, there are two especially important aspects that have a direct impact on the software system itself and the software models which we will see in the rest of the book: (i) business process models and (ii) enterprise architecture models.

3.4.1   BUSINESS PROCESS MODELING

Every organization has a myriad of business processes that manage all aspects of its daily operations. To put it simply, a business process is a set of related tasks that need to be performed to accomplish a certain goal (e.g., building a product or responding to a client complaint). These tasks may be executed in sequence, in parallel, waiting until a certain event occurs or a certain condition becomes true, etc.

When these processes are not formalized anywhere employees need to memorize them and managers must devote most of their time and attention to make sure processes are correctly followed. That's why making them explicit using some kind of precise language is a much better option for all companies. A business process model is the result of formalizing a business process. Beyond the control and clarification role, having this precise description of a business process enables all kinds of analysis regarding, for instance, the performance of the process opening the door to process reengineering opportunities to improve the overall organization.

Several business process modeling languages exist, but by far the most popular one is the *Business Process Modeling and Notation*[10] (BPMN) which is part of the family of OMG standard modeling languages (see Chapter 4). BPMN is a flow-like notation including graphical symbols to represents tasks/activities (the things to do), connections between those activities (how we move from one activity to the other), events (things that may happen during the flow), and lanes (to distinguish who is in charge of each part of the workflow) among others.

As an example, Figure 3.8 shows a simplified process model of the hiring process of an organization. Once the manager sets the job requirements, human resources is in charge of finding candidates. Potential candidates are notified and once they confirm by email (the process is stopped for that candidate until the email event is triggered) the actual interview takes place. An XOR gateway at the end of this interview task either notifies human resources to find another candidate or contacts accounting to setup the payroll depending on the result of the interview. Finally, the now new employee gets a first day briefing and the process ends.

---

[10]http://www.omg.org/spec/BPMN

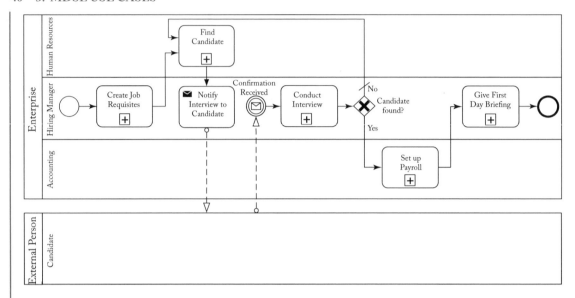

Figure 3.8: Example of a BPMN model.

The BPMN specification also provides a mapping between the BPMN notation and a workflow execution language, called Business Process Execution Language (BPEL). This helps to identify parts of the process model that may be automated with software components (e.g., proposing candidates based on the manager requirements if we already have a database of interesting people in the system) and/or link those activities with software systems already in place, monitoring the way that those systems behave as expected, i.e., behave as defined in the process model. Process mining techniques can be used to discover the actual process implemented by the software and compare it with the theoretical one.

3.4.2   ENTERPRISE ARCHITECTURE

Both the software perspective and the business process perspective are part of a broader concept that relates all aspects of an organization together, namely its Enterprise Architecture (EA). The EA can be defined as the analysis and documentation of an enterprise in its current and future states from a strategy, business, and technology perspectives where each perspective is interwoven with the rest. An EA helps the organization to manage change at all levels by considering first the impact of that change and its potential benefits across the organization.

Given its complexity, an EA cannot be represented using a single model but a combination of heterogeneous ones, organized as part of an architectural framework. Each company could define its own framework, but typically they use a reference framework that predefines a set of common principles, assumptions, and terminology that everybody can use to communicate better.

	Why	How	What	Who	Where	When
Contextual	Goal List	Process List	Material List	Organizational Unit and Role List	Geographical Locations List	Event List
Conceptual	Goal Relationship	Process Model	Entity Relationship Model	Organizational Unit and Role Relationship Model	Locations Model	Event Model
Logical	Rules Diagram	Process Diagram	Data Model Diagram	Role Relationship Diagram	Locations Diagram	Event Diagram
Physical	Rules Specification	Process Function Specification	Data Entity Specification	Role Specification	Location Specification	Event Specification
Detailed	Rules Details	Process Details	Data Details	Role Details	Location Details	Event Details

Figure 3.9: The Zachman framework for enterprise architecture.

Any framework provides a set of architecture views specifying the EA from the viewpoint of a specific stakeholder.

One of the first architectural frameworks, and for sure the most well-knwon, is the Zachman framework[11] developed in 1982 at IBM by John Zachman (but first published in 1987). It applies physics and basic engineering principles to the enterprise as a whole. It is not a methodology to develop your own EA but instead it provides a formal and structured way of viewing and defining your enterprise. The main perspectives in the Zachman framework are depicted in Figure 3.9 and are organized in a two dimensional table. Columns refer to the goal of that particular perspective, i.e., identify what is done there, how is it done, when, by whom, where, and why. Rows refer to the organization level this perspective refers to: contextual (executive level), conceptual (managerial level), logical (system's architect level), physical (engineering level), and detailed (user level). Some of the cells in the table will be covered in the book while others are perspectives at a higher abstraction level.

Even if Zachman is the classical reference framework, nowadays other frameworks like TOGAF[12] are more popular. In fact, many modeling tools that were originally UML-oriented are evolving to cover as well other EA perspectives defined in TOGAF to make the tool useful not only to software developers but to all stakeholders in the company.

Interestingly enough, this attempt to build modeling tools able to deal with the complexity of modeling full EAs, results in a series of challenges including the need to manage heteroge-

---

[11]http://www.zachman.com/about-the-zachman-framework
[12]http://pubs.opengroup.org/architecture/togaf9-doc/arch

neous models, their traceability and evolution, the definition of views mixing elements of different models, inter-model quality assurance, and many more. Some of them will be addressed in Chapter 10.

C H A P T E R   4

Model-driven Architecture (MDA)

The Object Management Group (OMG) has defined its own comprehensive proposal for applying MDE practices to systems development. This goes under the name of MDA (Model-Driven Architecture) [55]. We take MDA as a good exemplary MDE framework for two main reasons: first, MDA is a perfect case for explaining the MDE concepts introduced so far, as all the standard phases of a software development process such as analysis, design, and implementation are appropriately supported; second, given the importance of the OMG in the software industry, MDA is currently the most known modeling framework in industry. Without the aim of being complete, we report here some of the main contributions of the approach, which have become common knowledge and may ease the understanding of MDE in general.

MDA itself is not an OMG specification but rather an approach to system development which is enabled by existing OMG specifications, such as the Unified Modeling Language™ (UML®), the Meta Object Facility (MOF™), the UML Profiles SysML, SoaML, MARTE, and the CORBA® Component Model (CCM), among others.

The four principles that underlie the OMG's view of MDE are the following:

- Models must be expressed in a well-defined notation to foster effective communication and understanding of system descriptions for enterprise-scale solutions.

- Systems specifications must be organized around a set of models and associated transformations implementing mappings and relations between the models. All this enables a systematic design based on a multi-layered and multi-perspective architectural framework.

- Models must be built in compliance with a set of metamodels, thus facilitating meaningful integration and transformation among models and automation through tools.

- This modeling framework should increase acceptance and broad adoption of the MDE approach and should foster competition among modeling tool vendors.

Based on these principles, OMG has defined the comprehensive MDA framework and a plethora of modeling languages integrated within it.

4.1   MDA DEFINITIONS AND ASSUMPTIONS

The entire MDA infrastructure is based on a few core definitions and assumptions. The main elements of interest for MDA are the following.

- System: The subject of any MDA specification. It can be a program, a single computer system, some combination of parts of different systems, or a federation of systems.

- Model: Any representation of the system and/or its environment.

- Architecture: The specification of the parts and connectors of the system and the rules for the interactions of the parts using the connectors.

- Platform: A set of subsystems and technologies that provide a coherent set of functionalities oriented toward the achievement of a specified goal.

- Viewpoint: A description of a system that focuses on one or more particular concerns.

- View: A model of a system seen under a specific viewpoint.

- Transformation: The conversion of a model into another model.

As you can see, these official definitions are perfectly in line with the ones given in the introductory chapters on MDE and actually they represent the concrete incarnation and interpretation given by OMG to the various aspects of system definition. Based on these definitions, OMG proposes a whole modeling approach spanning methods, languages, transformations, and modeling levels. This chapter does not aim at providing a comprehensive coverage of the MDA framework. Instead, it is more aimed at giving an overview to the approach so as to let the reader understand the possibilities and potential usages.

4.2   THE MODELING LEVELS: CIM, PIM, PSM

As we introduced in Chapter 1, the level of abstraction of models can vary depending on the objectives of the models themselves. In MDA, three specific levels of abstraction have been defined, as shown in Figure 4.1.

- Computation-Independent Model (CIM): The most abstract modeling level, which represents the context, requirements, and purpose of the solution without any binding to computational implications. It presents exactly what the solution is expected to do, but hides all IT-related specifications, to remain independent of if and how a system will be (or currently is) implemented. The CIM is often referred to as a business model or domain model because it uses a vocabulary that is familiar to the subject matter experts (SMEs). In principle, parts of the CIM may not even map to a software-based implementation.

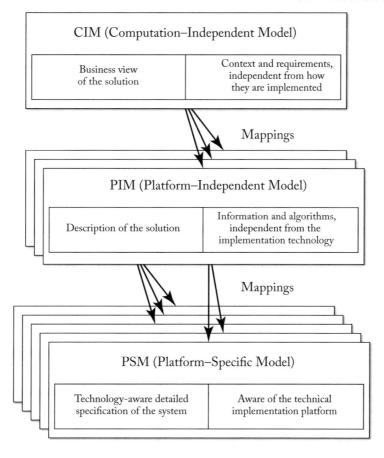

Figure 4.1: The three levels of modeling abstraction codified in MDA: computation-independent (CIM), platform-independent (PIM), and platform-specific (PSM) models.

- Platform-Independent Model (PIM): The level that describes the behavior and structure of the application, regardless of the implementation platform. Notice that the PIM is only for the part of the CIM that will be solved using a software-based solution and that refines it in terms of requirements for a software system. The PIM exhibits a sufficient degree of independence so as to enable its mapping to one or more concrete implementation platforms.

- Platform-Specific Model (PSM): Even if it is not executed itself, this model must contain all required information regarding the behavior and structure of an application on a specific platform that developers may use to implement the executable code.

A set of mappings between each level and the subsequent one can be defined through model transformations. Typically, every CIM can map to different PIMs, which in turn can map to different PSMs.

A concrete example of all three levels can be the following: a firm wants to implement an invoicing process. At the CIM level (Figure 4.2) the process is defined as a business process model listing the set of tasks to be performed and their dependencies. The PIM level will describe that the software application will do some parts of the job, in terms of information and behavior models (the access to a database and email sending could be fully automated, but shipping a signed paper invoice requires a clerk to mail the printed letter). Within this level, for instance, the details associated with a system implementation of the process must be detailed (e.g., Figure 4.3 shows a PIM description of the concept *Account*). Finally, at the PSM level, the implementation platform (whose description should be defined separately) for the process is selected and a set of precise descriptions of the technical details associated with that platform, including all the details regarding the invocations of the platform specific APIs and so on, will be provided within the models. For instance, Figure 4.4 shows the PSM description of the Account concept, including all the details of the specific implementation platform of choice (in this case, Enterprise Java Beans, EJB). The rationale is to keep at the PSM level all the details that are not relevant once the implementation moves to a different platform, thus keeping the PIM level general and agnostic.

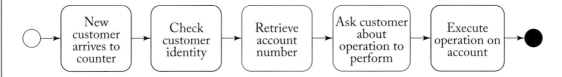

Figure 4.2: CIM example: the description of the business processes performed within an organization.

4.3   MAPPINGS

Generally speaking, a mapping consists of the definition of the correspondences between elements of two different models. Typically, a mapping is specified between two metamodels (i.e., modeling languages), in order to automate the transformations between the respective models. For instance, if the PIM level is described by a Petri Net, while the PSM is described by some UML dynamic diagram, the mapping must associate the concepts of Petri Nets with the corresponding ones in the UML diagram. Mappings between the metamodels allow us to build mappings between models, i.e., the traces (as described in Chapter 8). Mappings defined at the metamodel level are called *intensional*, while mappings defined at the model level are called *extensional* (because they are defined on the extension, i.e., instances, of the metamodel).

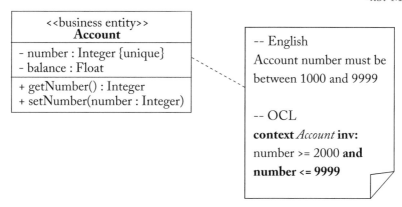

Figure 4.3: PIM example: the definition of a concept (*Account*), its properties, and constraints to be considered for any system implementation dealing with that concept.

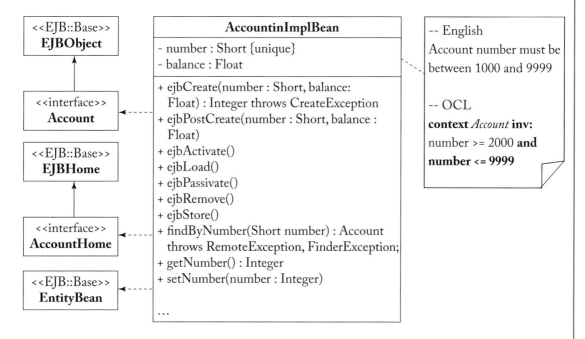

Figure 4.4: PSM example: EJB-based implementation model.

Mapping specifications can be defined in terms of: *weavings*, defining simple correspondences between metamodel elements; and *constraints*, describing the requirements for model transformations. From both specifications, model transformations may be derived.

The mapping gives rules and/or algorithms for the transformation of all instances of types in the metamodel of the PIM language into instances of types in the metamodel of the PSM language(s). Figure 4.5 illustrates this: the mapping from PIM to PSM is specified as a transformation defined between the platform-independent metamodel and the platform-dependent metamodel.

The mapping between different models (or modeling levels) can be implemented through transformations. Mappings can provide a conceptual specification for implementing the transformation between the different modeling levels. Obviously, the objective is to automate the mapping implementation as much as possible, thus avoiding expensive manual transformations.

In several cases, the mapping must take into account decisions that are taken along the development in terms of some details of the lower-level models. For instance, a mapping from PIM to PSM must take into account some decisions that apply to the PIM level based on choices at the PSM level. In these cases, a mapping defines annotations (*marks*, in MDA parlance) that will be used for guiding the transformations between the two levels. For instance, the transformation of a PIM to a PSM for the Enterprise Java Bean (EJB) platform will require a decision on which bean type should be created for the different PIM-level concepts. Basically, a mark represents a concept in the PSM which is applied to an element of the PIM in order to indicate how that element is to be transformed. A PIM plus all of the platform marks constitute the input to the transformation process resulting in a PSM.

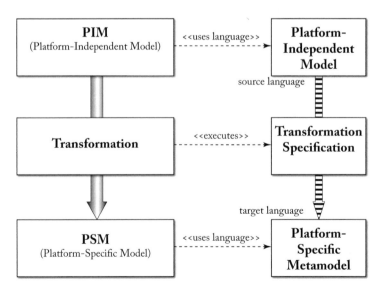

Figure 4.5: Representation of the transformation implementing a mapping between PIM and PSM levels.

4.4   GENERAL-PURPOSE AND DOMAIN-SPECIFIC LANGUAGES IN MDA

Besides the definition of the levels and mapping framework, the other main ingredients of MDA are the modeling languages used at each level. Indeed, one of the main OMG contributions is the standardization of a large number of modeling languages, spanning from general-purpose to domain-specific ones. Notice that the relationship between levels and modeling languages has cardinality M:N, in the sense that we can use more than one language to model the different aspects of a level, and vice versa a language can be used in more than one level.

In MDA, the core activity is the conceptualization phase in which analysis is conducted. First, requirements are codified as a PIM (or even CIM), and then the actual design of the application must be performed. This typically produces a PSM. In turn, the PSM is transformed into the running code, possibly requiring additional programming.

The cornerstone of the entire language infrastructure is the general-purpose language called UML (Unified Modeling Language), which is often perceived as a suite of languages, as it actually allows designers to specify their applications through a set of several different diagram types. Around and upon it, a set of domain-specific modeling languages (DSMLs) has been defined for addressing either vertical domains (e.g., finance, utilities, ecommerce) or horizontal domains (e.g., Web applications, mobile applications, service-oriented architectures), by exploiting either metamodeling techniques or the extensibility features provided within the UML language itself.

OMG supports UML Extensions through UML Profiles (as we will discuss in Chapter 6) and DSMLs by MOF. Typically, if you are dealing with object-, component-, or processes-oriented systems, you may reuse UML due to the already existing modeling support. If this is not sufficient, you can either use the inherent language extension mechanism of UML for a lightweight extension, or build a heavy-weight extension if the profile mechanism is not enough to tailor UML in the way you want, or build a completely new DSML from scratch. You may even use both, UML and DSMLs, on one level for different aspects or also for the same aspect (having a metamodel and a profile allows us to switch between UML and DSMLs) or on different levels. However, one should notice that adopting a design approach based on DSMLs is inherently different from using an approach based on UML (or any general-purpose language suite), as DSMLs will be definitely more tailored to the specific needs.

Notice that MDA is neutral with respect to this decision, in the sense that MDA is not favoring UML or DSMLs (although in the beginning of MDA, there was a strong focus on UML). In conclusion, MDA is about using CIM, PIM, and PSM: each of them can be defined either in plain UML, with a customized UML version, or using a DSML. Which languages you are employing, and if these languages are based on UML or not, is a different story.

4.5   ARCHITECTURE-DRIVEN MODERNIZATION (ADM)

Before closing the chapter on MDA, we also wish to mention another important initiative undertaken by OMG, that helps software developers in tackling the "reverse" problem, the problem of system evolution and modernization. In the same way that MDA is the OMG view of MDD, a corresponding initiative labeled ADM (Architecture-Driven Modernization) is addressing the problem of system reverse engineering and includes several standards that help on this matter. In particular, the Architecture-Driven Modernization task force of OMG [54] aims to create specifications and promotes industry consensus on modernization of existing applications, defined as any production-enabled software, regardless of the platform it runs on, language it is written in, or length of time it has been in production. The ADM taskforce aims at improving the process of understanding and evolving existing software applications that have entered the maintenance and evolution mode, so as to grant: software improvement, modifications, interoperability, refactoring and restructuring, reuse, migration, service-oriented integration, and so on.

The final aim of the initiative is to provide conceptual tools for reducing the maintenance effort and cost, and for extending the lifecycle of existing applications by enabling viable revitalization of existing applications through a model-based approach.

The ADM taskforce has defined a set of metamodels that allow describing various aspects of the modernization problem.

- The Knowledge Discovery Metamodel (KDM): An intermediate representation for existing software systems that defines common metadata required for deep semantic integration of lifecycle management tools. KDM uses OMG's MOF to define an XMI interchange format between tools that work with existing software as well as a general-purpose, abstract interface for modernization tools. KDM supports knowledge discovery in software engineering artifacts.

- The Software Measurement Metamodel (SMM): A meta-model for representing measurement information related to software, its operation, and its design. The specification is an extensible meta-model for exchanging software-related measurement information concerning existing software assets (designs, implementations, or operations).

- The Abstract Syntax Tree Metamodel (ASTM): A complementary modeling specification with respect to KDM. While KDM establishes a specification for abstract semantic graph models, ASTM establishes a specification for abstract syntax tree models. Thus, ASTM supports a direct mapping of all code-level software language statements into low-level software models. This mapping is intended to provide a framework for invertible model representation of code written in any software language (achieved by supplementation of the ASTM with concrete syntax specifications).

This set of metamodels is a valid instrument for enabling the typical model-driven reverse engineering (MDRE) and modernization scenarios also presented in Chapter 3 and reported in

Figure 4.6. Starting from the PSM of the system as-is, designers extract the PIM models through knowledge discovery, possibly exploiting 1-to-1 mappings to implementation language models as offered by ASTM. Then they work for making these models compliant with the requested compliance levels (e.g., described according to SMM) with respect to functional and non-functional requirements (possibly given at CIM level). Finally, they transform the system and process into seamless MDA-compliant solutions for the design and implementation of the to-be version.

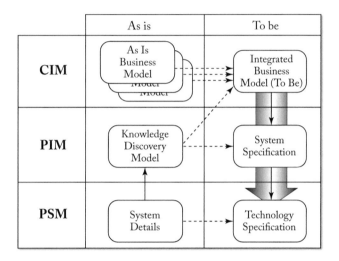

Figure 4.6: Overview of a typical MDRE scenario, as supported by ADM tools.

CHAPTER 5

Integration of MDSE in your Development Process

MDSE is process-agnostic, i.e., it neither provides nor enforces any specific development process, but it can be integrated in any of them. Furthermore, MDSE *per se* does not define which models must be used in each step of the development process, at what abstraction level, how they must be related, and so on. It is up to each organization to define the right methodology (*right* in terms of the organization's context and requirements) to successfully apply MDSE.

Nevertheless, as happens with the introduction of any complex technology in an organization, adoption of MDSE is not straightforward. In fact, one of the typical failure reasons for MDSE projects was to think that you build a code generator and start increasing your productivity tenfold the day after. Integration of MDSE must be a step-by-step process that takes into account your specific organizational context.

This chapter first provides some general considerations on the adoption of MDSE and then discusses how MDSE can be merged with five well-known software development approaches of various kinds.

5.1 INTRODUCING MDSE IN YOUR SOFTWARE DEVELOPMENT PROCESS

We have witnessed how some organizations interested in giving MDSE a try drop it after an initially failed attempt or are unable to draw all the benefits that supposedly come with the use of MDSE in practice.

As with any other technology adoption, organizational, managerial, and social aspects (as opposed to simply technical ones) are the main factors of failure [34]. Neglecting to consider these aspects hampers the successful adoption and efficient application of a new MDSE-based process, eluding the benefits that modeling could bring, and even worse, discouraging its use in future projects.

Some aspects to take into account in your first MDSE project are common sense: start with a small project, choose a non-critical one (so that there is no additional pressure for the development team), make sure you have the commitment of the management to support you when inevitably some problems arise during the project, get some external help (e.g., an expert MDSE consultant to help you iron out the details of the adoption process), and so forth.

Nevertheless, other aspects are a little bit more subtle, and therefore, deserve more careful attention. In the following, we comment on what we believe are two of the major challenges in the adoption of MDSE: (i) the imbalanced distribution of efforts and rewards among team members, and (ii) the lack of socio-technical congruence.

5.1.1   PAINS AND GAINS OF SOFTWARE MODELING

The adoption of a model-based process introduces new tasks and roles into the development process. These additional tasks are expected to be beneficial for the quality and productivity of the overall team. However, at the personal level, not all members benefit equally from the change. In fact, some members, depending on their assigned role, can perceive modeling as detrimental since it imposes additional burdens that benefit other people in the team but not themselves. Therefore, they may not see MDSE as cost effective: they suffer the cost but they may not realize the positive overall impact.

As an example, when precise and up-to-date UML models are available, correctness of changes performed during the software maintenance is clearly improved. This will benefit the subteam in charge of that task, but the chore of developing those models would fall on the analysis subteam responsible for creating such models. When both activities are performed by the same team members, they easily realize that the extra effort of creating the UML models (the "pain") at the beginning pays off in the end since they benefit from using those same models (the "gains") when performing maintenance tasks. However, when the activities are performed by two different groups, the group in charge of doing the models does not directly benefit from those models and may doubt that their effort is worthwhile.[1]

Therefore, when the work distribution among the members separates the pains and gains of modeling, we must adopt policies that recognize (and compensate for) those differences. Otherwise, we may face motivation problems during the performance of the pain activities which would limit the benefits we may get with the gain ones.

5.1.2   SOCIO-TECHNICAL CONGRUENCE OF THE DEVELOPMENT PROCESS

The technical requirements of the MDSE-based process may disrupt the social elements of the team due to the introduction of new roles, skills, coordination patterns, and dependencies among the team members.

All these elements must be carefully considered to ensure that the organization fits the socio-technical requirements of the new MDSE process, where fit is defined as the match between a particular design (social structure) and the organization's ability to carry out a task under that design [16]. For instance, the organization must be able to match the roles of the target process

---

[1]The difficulties of introducing an application/process that provides a collective benefit but not a uniform individual benefit is well known, e.g., see [31] for a study for the groupware domain.

on the actual team members, ensuring that each member has the skills required to play that role. A good socio-technical congruence is a key element in the software development performance.

As an example, imagine the extreme case where an organization with no modeling experience decides to adopt a full MDD approach with complete automatic code generation. This change has a lot of (direct and indirect) effects on the development team and its social organization. Therefore, to succeed in this endeavor, the team needs to be willing and able to adopt new roles (e.g., the platform expert, language engineer, transformation developer, domain expert) with the corresponding new skills (e.g., proficiency in modeling, model transformation, and language development techniques) and dependencies (the designer depends on the platform expert for knowing the platform capabilities, on the language engineer for the language to use when defining platform-specific models, on the transformation developer for transforming platform-independent models into platform specific ones, etc.). This is not a trivial evolution and requires time and means (e.g., to take training courses) but it is a necessary condition for the successful adoption of MDSE.

5.2   TRADITIONAL DEVELOPMENT PROCESSES AND MDSE

Under the banner of "traditional" development processes, we group all development processes that prescribe to follow the sequence of "classical" phases (requirement elicitation, analysis, design, implementation, maintenance, etc.) in any software development project, regardless whether these activities are performed following a waterfall model, spiral model, an iterative and incremental model (like in the well-known Unified Process [37]), etc. This is in contrast with the development processes described in the next sections which depart significantly from this schema.

In general, all traditional processes are already model-based, i.e., they consider models as an important part of the process and propose to use different kinds of models to represent the system in the successive stages of the development process. In this sense, most include the notion of requirement models, analysis models, and design models. What MDSE brings to these processes is the possibility of going from *model-based* to *model-driven*, i.e., the opportunity to use MDSE techniques to (at least partially) automate the transition between the different phases of the development process. Model-to-model transformations may be used to refine the models in the early phases, while model-to-text transformations may be employed to generate the actual software implementation at the end of the process.

5.3   AGILE AND MDSE

Agile methods are a family of development processes that follow the principles of the Agile Manifesto,[2] like Scrum, XP, and others. The Agile Manifesto proposes to center the development around four main principles:

---

[2]http://agilemanifesto.org

1. individuals and interactions over processes and tools;

2. working software over comprehensive documentation;

3. customer collaboration over contract negotiation; and

4. responding to change over following a plan.

These values are implemented in a dozen principles related to the frequent delivery of working software, tight co-operation between customers and developers, continuous attention to the good design of the software, and so forth.

So far, Agile and MDSE have not often been used in combination. The Agile community tends to see MDSE as a superfluous activity that has no place in their view of *working software* as the only measure of project progress. Their main criticisms of MDSE (summarized by Stephen Mellor) are that models:

- don't run (i.e., they are not working software);

- can't be tested;

- are just documentation; and

- require extra work and alignment (i.e., they are not adequate for late requirement changes).

However, these criticisms are basically resulting from a limited understanding of the concept of model. If one interprets *models as sketches* (you draw them on your whiteboard and then you just throw them away) or *models as blueprints* (aimed at directing and planning the implementation before actually building it) then it is true that this view of modeling is not agile.

However, we have already seen in Chapter 3 that *models can be executable artifacts*. In this case, models are neither just pictures nor guidelines. They are meant to be an active part of the system's implementation and verification. They are built under the assumption that design is less costly than construction, and thus, it makes sense to put an effort on their realization, because then construction will come (almost) for free. In this sense, models are working software. Therefore, you can definitely be agile while developing with model-driven techniques.

Some initial proposals show how these two paradigms can interact. The most relevant is the *Agile Modeling* (AM) [4] initiative lead by Scott W. Ambler.[3] Simply put, AM is a collection of modeling practices that can be applied to a software development project in an effective and light-weight (i.e., agile) manner. The goal is to avoid "modeling just for the sake of modeling." The principles of AM can be summarized in the following:

- *Model with a purpose*. Identify a valid purpose for creating a model and the audience for that model.

---

[3]http://www.agilemodeling.com

- *Travel light.* Every artifact will need to be maintained over time. Trade-off agility for convenience of having that information available to your team in an abstract manner.

- *Multiple models.* You need to use multiple models to develop software because each model describes a single aspect/view of your software.

- *Rapid feedback.* By working with other people on a model you are obtaining near-instant feedback on your ideas.

- *Favor executable specifications.*

- *Embrace change.* Requirements evolve over time and so do your models.

- *Incremental change.* Develop good enough models. Evolve models over time in an incremental manner.

- *Working software is your primary goal.* Any (modeling) activity that does not directly contribute to this goal should be questioned.

- *Enabling the next effort is your secondary goal.* Create just enough documentation and supporting materials so that the people playing the next game can be effective.

These agile modeling practices and principles are the basis of what we could call an *Agile MDD* approach where the focus is on the effective modeling of executable models to transition from models to working software automatically in the most agile possible way.

Note that not only can we integrate MDSE practices in an agile development process but the inverse is also true. We can reuse agile techniques to improve the way we perform MDSE and build MDSE tools. Why not apply eXtreme Programming (XP) and other agile techniques when designing new modeling languages, models, or model transformations? If so many studies have proven the benefits of eXtreme Programming, why not have eXtreme Modeling as well?

5.4   DOMAIN-DRIVEN DESIGN AND MDSE

Domain-driven design (DDD) [22] is an approach for software development based on two main principles:

1. The primary focus of a software project should be the domain itself and not the technical details.

2. Complex domain designs should be based on a model.

As such, DDD emphasizes the importance of an appropriate and effective representation of the problem domain of the system-to-be. For this purpose, DDD provides an extensive set of design practices and techniques aimed at helping software developers and domain experts to share and represent with models their knowledge of the domain.

Clearly, DDD shares many aspects with MDSE. Both argue the need of using models to represent the knowledge of the domain and the importance of first focusing on platform-independent aspects (using the MDA terminology) during the development process. In this sense, MDSE can be regarded as a framework that provides the techniques (to model the domain, create DSLs that facilitate the communication between domain experts and developers if needed, etc.) to put DDD in practice.

At the same time, MDSE complements DDD (Figure 5.1) by helping developers to benefit even more from the domain models. Thanks to the model transformation and code generation techniques of MDSE, the domain model can be used not only to represent the domain (structure, rules, dynamics, etc.) but also to generate the actual software system that will be used to manage it.

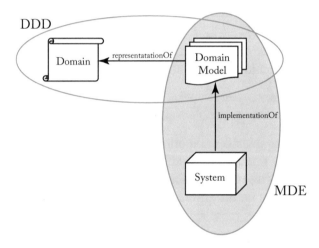

Figure 5.1: Relationship between DDD and MDD.

5.5   TEST-DRIVEN DEVELOPMENT AND MDSE

Test-driven development (TDD) [8] relies on the test-first philosophy. This means, the developer first writes an executable test case that will be used to check the new functionality to be implemented. If the test fails (which is supposed to happen since the new functionality has not yet been coded), the developer writes the code needed to pass the test (i.e., which in fact "forces" to implement the functionality). Once the new code passes the full test suite, the code can be refactored and the process starts again.

MDSE can be integrated into a TDD process at two different levels, depending on whether models will be used as part of a code generation strategy to automatically derive the system implementation or not.

5.5.1   MODEL-DRIVEN TESTING

When models are used to specify the system but the system is not automatically generated from them, models can be used to derive the tests that the system's implementation will need to pass in order to ensure that it behaves as expected (i.e., it behaves as defined in the model). This is known as model-based testing. As shown in Figure 5.2, a test generation strategy is applied at the model level to derive a set of tests. This set is subsequently translated into executable tests that can be run on the technological platform used to implement the system. Testing strategies depend on the kind of model and test to be generated. For instance, constraint solvers can be used to generate test cases for static models while model checkers can be used to generate relevant execution traces when testing dynamic models.

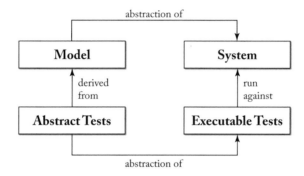

Figure 5.2: Model-based testing.

5.5.2   TEST-DRIVEN MODELING

When software is derived from models then there is no need to test the code (of course, assuming that our code generation is complete and that we trust the code generator). Instead, we should test the models themselves. In this sense, some approaches for the test-driven development of modeling artifacts have been proposed, e.g., [69]. These approaches follow the same philosophy: before developing a model excerpt, the modeler needs to write the model test that will be used to evaluate the new functionality at the modeling level. Then this new functionality is modeled and the test is executed to ensure that the model is correct. This practice requires models to be executable (see Chapter 3). Chapter 10 discusses some tools that help in the exhaustive and systematic testing of models and/or in their validation.

5.6   SOFTWARE PRODUCT LINES AND MDSE

Software product lines [58] are becoming mainstream for the systematic development of similar systems. The main idea of software product lines is to design a family of systems instead of one single system. A family of systems defines a set of systems that share substantial commonalities but also exhibit explicit variations. In this respect, productivity and quality improvements achieved by software product lines originate from the reuse of so-called core assets, i.e., elements shared by all systems of a family. In a nutshell, software product lines can be seen as a course-grained software reuse mechanism which allows us to reuse the common parts but also manages variations between the systems of a family. In order to do so, software product line engineering makes a clear distinction between domain engineering, i.e., the development of reusable core assets within a domain, and application engineering, i.e., assembling of core assets for building the actual systems.

Model-driven engineering techniques have proven to be effective for supporting software product line engineering [38]. One of the most interesting question when it comes to software product lines and model-driven engineering is how to formulate the knowledge of a software product line by using models. Especially, the variability aspect is in this context of main interest as modeling languages are quite often lacking variability as a first-class language concept. In the past years, a plethora of variability modeling approaches have been presented to elevate this shortcoming. In general, two types of modeling approaches can be considered: annotative and compositional [38]. Annotative modeling approaches aim to represent all elements of a system family within one model description, and in addition explicit activation or deactivation steps are needed for representing a concrete system. In contrast, compositional modeling approaches are using a set of model fragments which have to be combined to represent a concrete system.

One of the most promising combinations of software product lines and model-driven engineering has been proposed within the scope of feature-oriented software development (FOSD). The main idea of FOSD is to decompose a family of software systems in terms of its features [5]. Features may be considered as visible aspects of the software system from a user's viewpoint or more generally as characteristics of the domain. From a set of features of a software system family, the systems can be generated by selecting certain features.

One approach to specify a feature-based decomposition of a system family is the usage of feature models. Feature models express the commonalities and variabilities among the systems of a software product line. These models organize the so-called features in a hierarchical structure. In this context, the basic relationships between a feature and its children are: mandatory relationships, optional relationships as well as XOR, OR, and AND groupings. In addition, cross-tree relationships may be introduced for describing inclusion/exclusion relationships of features residing in different parts of the feature model. By selecting certain features, a configuration or concrete system can be produced. As a feature model is a formal model in the sense that the relationship types clearly define which combinations of feature selections are valid or invalid, there is a systematic derivation of configurations supported. Furthermore, this supports the completion of a partial configuration, i.e., a partial selection of features which should be in a system, by reasoning which features are needed in addition to the selected ones to end up with a valid configuration.

An example feature model is shown in Figure 5.3. In particular, this model defines an abstract system consisting of different A-D features. Please note that there is a mandatory A feature which has to be selected for any system, but there is also an optional D feature which can be chosen independently of any other feature. Furthermore, for some features all children may be selected (OR) or only one of them may be selected (XOR). Finally, there is also a cross-tree constraint shown which states: if feature B1 is selected then feature C1 has to be selected as well.

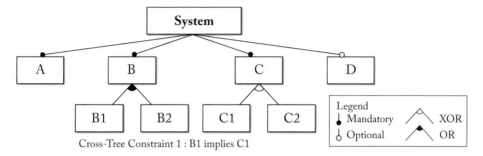

Figure 5.3: Example feature model.

Besides feature models, a plethora of other modeling languages for modeling software product lines have been proposed. In addition, several model-based techniques for the automated processing of variability models, including the automated analysis of variability models, derivation of configurations from the variability model of a software product line, and generation of test cases for derived products have been developed. Thus, the combination of model-based techniques and software product line engineering constitutes a promising reusability approach for software systems. Not surprisingly, this has been already explored in the past. For instance, Software Factories [30] combine domain-specific modeling and FOSD to exploit a structured collection of related software assets and processes that include reusable code components, documentation, and reference implementations to build up new applications.

CHAPTER 6

Modeling Languages at a Glance

Modeling languages are conceptual tools with the purpose of letting designers formalize their thoughts and conceptualize the reality in explicit form, be it textual or graphical. This chapter describes the main features of modeling languages, considering the peculiarities of general-purpose languages (GPLs), domain-specific languages (DSLs), and the intermediate solutions that allow the customization of GPLs for specific purposes.

6.1 ANATOMY OF MODELING LANGUAGES

A modeling language is defined by the following three core ingredients:

- *Abstract syntax*: Describing the structure of the language and the way different primitives can be combined, independently of any particular representation or encoding.

- *Concrete syntax*: Describing specific representations of the modeling language, covering encoding and/or visual appearance issues. The concrete syntax can be either textual or graphical. The concrete syntax is what the designer usually looks up as a reference in modeling activities. If the syntax is a visual one, the result of the modeling activity consists of one or more diagrams.

- *Semantics*: Describing the meaning of the elements defined in the language and the meaning of the different ways of combining them.

These three ingredients are mandatory for a modeling language, which is not well defined if any of the three is missing or incomplete. Figure 6.1 shows the three ingredients and their relationships. The semantics define the meaning of abstract syntax and, indirectly, of concrete syntax; the concrete syntax represents the abstract syntax.

This applies both to general-purpose languages (GPLs) and domain-specific ones (DSLs). Unfortunately, both language designers and users tend to forget about some aspects. In particular, the semantics is often neglected in the definition or in the usage of the language. This is probably due to the fact that the concrete syntax is the most visible part of the language, as it represents the actual notation which is used in the everyday application of the language.

However, it's important to point out that it doesn't make sense to define a language without fully specifying the conceptual elements that constitute it and their detailed meaning. A partial or

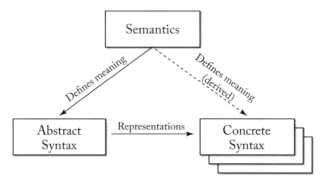

Figure 6.1: The three main ingredients of a modeling language (semantics, abstract syntax, and concrete syntax) and their relationships.

wrong specification of the semantics paves the way to wrong usages of the language and to misunderstandings of the meaning of language elements and purpose. This also leads to situations where different people understand differently the concepts and models, due to different interpretations of the language symbols. That's why a lack of appropriate understanding of a language's semantics is extremely dangerous.

As discussed in Chapter 1, modeling languages are used for describing the real world at a certain level of abstraction and from a specific point of view. The semantics of the language aims at describing this in detail and enables the correct usage of the language. The semantics of a language can be defined as:

- *Denotational semantics*: by defining meaning of all the concepts, properties, relationships, and constraints through mathematical expressions.

- *Operational semantics*: by defining the meaning of the language by implementing an interpreter that directly defines the model behavior.

- *Translational semantics*: by mapping the language concepts to another language with clearly defined semantics.

6.2   MULTI-VIEW MODELING AND LANGUAGE EXTENSIBILITY

In software engineering, you typically need to model orthogonal aspects of a system. This can be achieved either by combining different modeling languages (each one focusing on one specific perspective) or by using a multi-viewpoint modeling language, including different diagram types.

The system aspects to be described are often classified into two main categories: *static* (or *structural*) aspects and *dynamic* aspects. The former describes the main ingredients of the modeled entity and their relations, while the latter describes the behavior of such ingredients in terms of actions, events, and interactions that may occur between them. Keeping static and dynamic aspects separated is usually a good practice.

Furthermore, the languages often include some extensibility mechanisms that allow designers to expand the coverage of the language by defining new modeling elements. This is very common in GPLs, which often need some specialization when used for specific domains or problems. Sometimes this leads to a sort of well-known domain-specific language developed out of the specialization of general-purpose ones.

A well-known example of the scenario described above is UML, the GPL proposed by OMG within the MDA framework. The language is a general-purpose one and it comprises a wide set of different diagrams, which can be used together for describing a system, and a set of extension mechanisms. Several extensions have been proposed, and some of them in turn have become standard languages, such as SysML, SoaML, and others. Therefore, UML can be seen as a *family of languages* (or *language suite*) in three senses. First, UML as a whole (considering all the symbols and diagram types) allows enough variability to be considered a group of languages. Second, it pairs with a set of correlated languages such as OCL that improve the expressive power of UML itself. Third, UML is accompanied by a set of domain-specific profiles defined around it.

6.3   GENERAL-PURPOSE VS. DOMAIN-SPECIFIC MODELING LANGUAGES

As mentioned in Chapter 1, one of the main classification of modeling languages concerns the distinction between domain-specific and general-purpose languages.

Domain-Specific Modeling Languages (*DSMLs*, or *DSLs* for short)[1] are languages that are designed on purpose for a specific domain, context, or company, to support people who need to describe things in that domain.

In contrast, *General-Purpose Modeling Languages* (*GPMLs*, or *GMLs/GPLs*)[2] represent modeling notations that can be applied to any sector or domain for modeling purposes.

However, we want to highlight that this distinction is not so deterministic and well defined since deciding whether a language is a DSL or GPL is not a binary choice. For instance, at first sight UML and similar languages seem to belong to the latter group in this classification since UML can be used to model any domain. However, if we consider modeling in general, we can see UML as a DSL tailored to the specification of (mainly object-oriented) software systems.

Moreover, the decision on whether to use UML (or any other GPL) or a DSL for your next modeling project is far from a trivial one. It's clear that for some domains (for instance, user interaction design) UML is not exactly the best option and that a domain-specific language

---

[1]In the modeling realm, domain-specific languages (DSLs) are sometimes called DSMLs just to emphasize that they are modeling DSLs. However, DSL is by far the most adopted acronym and therefore we will stick to it in the upcoming discussions.
[2]Same discussions of DSL vs. DSML applies to GPML. We will use GPL for simplicity.

tailored to that domain may produce much better results. Nevertheless, the discussion DSL vs. UML leads nowhere if it is just UML bashing. We are all well aware of the negative aspects of UML, but what we should never forget is that UML may not be a domain-specific language but it is a many-domains language (MDL).[3] That is, UML may not be suited for all domains but can be directly applied to easily and successfully model many of them.

UML may have many defects, but at least after all these years we are more or less aware of all of them. Instead, when creating an unnecessary DSL you may end up repeating many of the mistakes that we have managed to solve (or at least identify) in the case of UML. Creating DSLs that just reinvent the wheel again and again should be discouraged at all costs. If your DSL resembles UML too much maybe you should consider using just (a subset of) UML and avoid inventing new "almost-UML" notations and considering them as new DSLs.

On the theoretical side, concerning GPLs, one may wonder whether a language to be a GPL must also be a Turing-complete language. The question is a valid one, because, if not, it means that the language is not really "general," since it cannot solve each problem. However, this interpretation is typically not considered at all, and the definition of a language as GPL or DSL is more a matter of practice or subjectivity. Martin Fowler clarifies that even just the decision on whether a set of concepts (or operations) is actually a language or just a set of operations within another language is a matter of perspective[4]:

> *"As with the internal DSL vs API issue, I think the intention is the key here, both of the language writer and of the user. If I use XSLT only to transform an XML document then I'm treating it as a DSL - even if I do some fancy call out to embed some value in the output document. However, if I use it to solve the eight queens puzzle I'm using it as a GPL. If the designers of XSLT see it as about XML transformations then they are thinking of it as a DSL even if clever people do unnatural acts with it."*
> (Martin Fowler)

6.4   GENERAL-PURPOSE MODELING: THE CASE OF UML

This section aims at providing an overview on the *Unified Modeling Language* (*UML*).[5] Besides being widely known and adopted, UML as an example language is interesting for discussing the features and aspects presented so far as general characteristics for modeling languages. Indeed, UML is a full-fledged language suite, since it comprises a set of different diagrams for describing a system from different perspectives. Figure 6.2 shows the taxonomy of UML diagrams. There are 7 different diagrams which can be used for describing the static (structural) aspects of the system, while another 7 diagrams can be used for describing the dynamic (behavioral) aspects. As shown in Figure 6.3, some of these diagrams describe characteristics of classes (i.e., abstract concepts), while others are used for describing features and behavior of individual items (i.e., instances). Some diagrams can be used for describing both levels.

---

[3]http://modeling-languages.com/uml-mdl-many-domains-language
[4]http://martinfowler.com/bliki/DslBoundary.html
[5]The last official version of UML can always be retrieved at: http://www.omg.org/spec/UML.

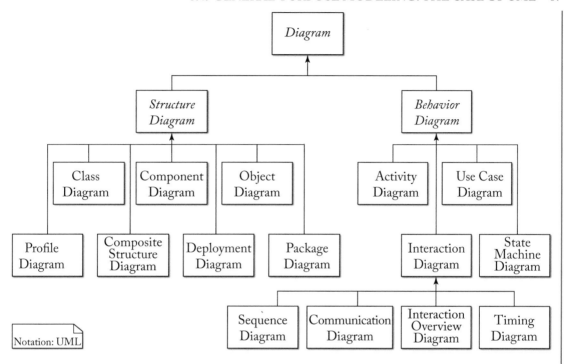

Figure 6.2: UML diagrams taxonomy.

Notice that UML is just a modeling language, and does not enforce any particular development method. UML is often mentioned together with the Unified Process (UP), best known in its specialization called Rational Unified Process (RUP). RUP is an iterative software development process framework created by Rational Software Corporation (now a division of IBM). While RUP and UML work very well together, the adoption of RUP is not necessary for using UML.

UML has a long history which comprises merges of several pre-existing approaches and subsequent restructuring of its specifications. Some pieces of UML, such as Harel's statecharts and Booch's notations, date back to the mid-1980s. At that time, a plethora of different notations existed. However, the first version of UML was conceived only in 1996, born from the fusion of the approaches of Booch, Rumbaugh, and Jacobson. This proposal was discussed at the OOPSLA conference in 1996. Other integrations followed and led to the submission of UML to OMG in August 1997 and the official release as an OMG standard in November 1997 (as UML 1.1). The long and controversial history of UML is accountable for the complexity and also some inconsistencies of the language that hold still today. However, despite some important and motivated criticisms to the language, UML remains a unifying language for software modeling.

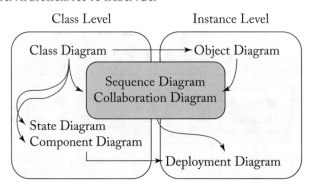

Figure 6.3: Dichotomy between class- and instance-based models in UML.

6.4.1   DESIGN PRACTICES

UML provides a set of facilities for designing systems and enables good design practices, including the following ones:

- Using several integrated and orthogonal models together: UML comprises a suite of diagrams that share some symbols and allow cross-referencing between modeling artifacts.

- Modeling at different levels of detail: UML allows eliding details in diagrams if needed, so as to allow designers to choose the right quantity of information to include in diagrams, depending on the purpose of the modeling activity and on the phase of the development process.

- Extensibility: UML provides a good set of extensibility features which allow us to design customized modeling languages if needed.

- Pattern-based design: A set of very well-known design patterns [27], defined by the so-called Gang of Four, is available.

  The following sections delve into the description of a few details about UML modeling, so as to provide an overview on how a typical GPL would look.

6.4.2   STRUCTURE DIAGRAMS (OR STATIC DIAGRAMS)

Structure diagrams emphasize the description of the elements that must be present in the system being modeled. They are used extensively in documenting the software systems at two main levels.

- The *conceptual items* of interest for the system. This level of design aims at providing a description of the domain and of the system in terms of concepts and their associations. This part is typically described by the following diagrams.

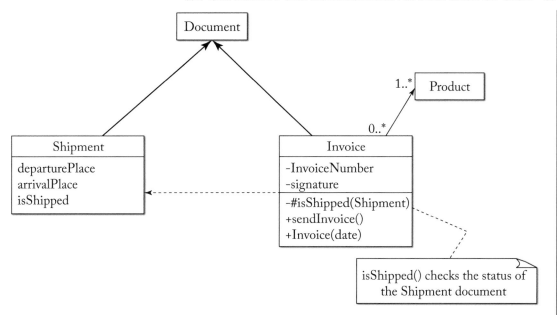

Figure 6.4: UML class diagram example.

1. *Class diagram*: Describes the structure of a system by classes, their attributes, and the relationships among the classes. A class diagram example is shown in Figure 6.4. The class Document (shown elided, i.e., without all the details on attributes and methods) is specialized by the two sub-classes, Shipment and Invoice. Notice that classes can be described at different levels of detail: Document is shown in the elided form, with only the class name; Shipment also shows its attributes; while Invoice shows both attributes and methods. Invoice has a dependency on Shipment, because the method isShipped is checking the value of the isShipped attribute of Shipment, as described by the comment. Invoice is connected to Product through an association, whose multiplicity is also specified.

2. *Composite structure diagram*: Describes the internal structure of a class and the collaborations that this structure enables. The syntax is very similar to that of class diagrams, but it exploits containment and connections between items. The core concepts that are used in these diagrams are *parts* (i.e., roles played at runtime by one or more instances), *ports* (i.e., interaction points used to connect classifiers with each other), and *connectors* (represented as undirected edges that connect two or more entities, typically through ports).

3. *Object diagram*: A view on the structure of example instances of modeled concepts at a specific point in time, possibly including property values. Again, the syntax is similar

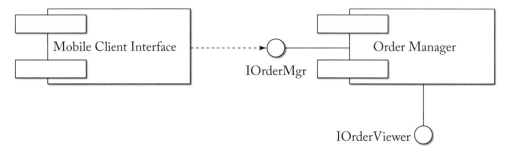

Figure 6.5: UML component diagram example.

to that of class diagrams, but objects are identified by underlined names. Objects can be identified by the tuple object name and class name, by the simple object name, or by the class's name they are instantiated from. The respective notations are as follows:

objectName:ClassName or objectName or :ClassName

- The *architectural representation* of the system. This level of design aims at providing a description of the architectural organization and structure of the system. It typically consists of reuse and deployment-oriented information, which aggregates or bases on the conceptual modeling done in the previous step. The diagrams involved in this part of the design comprise the following:

  1. *Component diagram*: Describes how a software system is divided up into components and shows the dependencies among these components. A component in UML is defined as a distributable piece of implementation of a system, including software code (source, binary, or executable) but also other kinds of information. A component can be thought of as an implementation of a subsystem. Figure 6.5 shows a UML component diagram example, where two components interact by exposing an interface and declaring a dependency on it.

  2. *Package diagram*: Describes how a system is divided into logical groupings, called packages in UML parlance, by showing the dependencies among these groupings.

  3. *Deployment diagram*: Describes the way in which the software artifacts are deployed on the hardware used in systems' implementations.

6.4.3   BEHAVIOR DIAGRAMS (OR DYNAMIC DIAGRAMS)

Behavior diagrams describe the events and interactions that must happen in the system being modeled. Different diagrams are available for describing the dynamic aspects and some of them are equivalent in terms of carried information. Depending on which are the more prominent aspects that the designer wants to highlight, one may opt for one or another graphical notation.

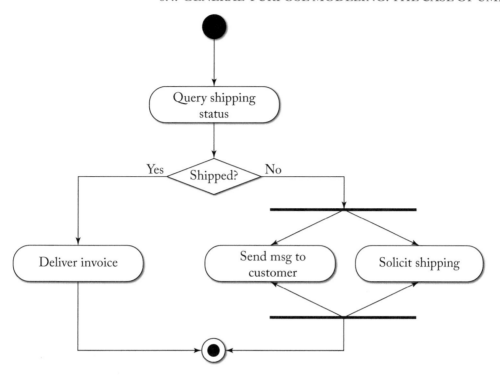

Figure 6.6: UML activity diagram example.

Behavioral models designed with these diagrams usually do not describe the whole system at once. Every model describes a single or a few features of the system and the dynamic interactions they involve. The behavioral diagrams are the following:

- *Use case diagram*: Describes the functionality provided by a system in terms of actors external to the system, their goals in using the system, represented as use cases, and any dependencies among those use cases. This diagram is very useful for understanding the borders of the system to be implemented and for clarifying the usage scenarios. Furthermore, it is in particular relevant in the requirement specification phase and early design phase of the software development process.

- *Activity diagram*: Describes the step-by-step workflows of activities to be performed in a system for reaching a specific goal. An activity diagram shows the overall flow of data and control for performing the task, through an oriented graph where nodes represent the *activities*. Figure 6.6 shows a UML activity diagram example, describing the actions to undertake depending on the status of the Shipping of a product.

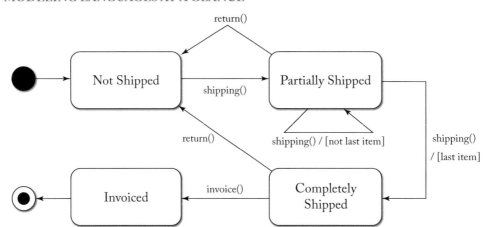

Figure 6.7: UML state diagram example.

- *State machine diagram* (or *statechart*): Describes the states and state transitions of the system, of a subsystem, or of a specific object. In general, state machines are suitable for describing event-driven, discrete behavior, but are inappropriate for modeling continuous behavior. The example in Figure 6.7 shows the states of a product with respect to the shipping and invoicing aspects, and the transitions that lead from one state to another. Transitions may have associated conditions (e.g., whether the item is the last of the shipment or not). Please notice that in statecharts the nodes represent states (while in activity diagrams they represented actions).

- *Interaction diagrams*: A subset of behavior diagrams which emphasizes the flow of control and data among the elements of the system. This set comprises the following diagrams.

  1. *Sequence diagram*: Shows how objects communicate with each other in terms of a temporal sequence of messages. The time dimension is the most visible aspect in these kinds of diagrams, as messages are sequenced according to a vertical timeline, and also the lifespan of objects associated with those messages is reported. Figure 6.8 shows an example sequence diagram describing three objects (instances of classes Shipment, Invoice, and Product) and the messages exchanged between each other for printing an invoice. Interaction diagrams describe execution scenarios of the system.

  2. *Communication or collaboration diagram*: Shows the interactions between objects or classes in terms of *links* (represented as solid undirected lines connecting the elements that can interact) and *messages* that flow through the links. The sequence information is obtained by numbering the messages. They describe at the same time some static structure (links and nodes) and dynamic behavior (messages) of the system. They represent

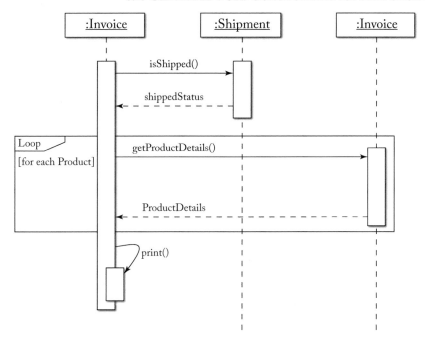

Figure 6.8: UML sequence diagram example.

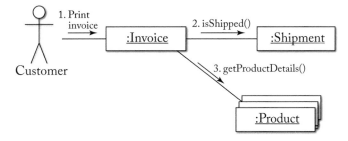

Figure 6.9: UML collaboration diagram example.

a combination of information taken from class, sequence, and use case diagrams. An example is shown in Figure 6.9.

3. *Interaction overview diagram*: Provides an overview in which the nodes represent interaction diagrams.

4. *Timing diagrams*: A specific type of interaction diagram where the focus is on timing constraints.

6.4.4   UML TOOLS

Given the popularity of UML, a wide set of UML modeling tools is available. They constitute a massive category of modeling tools in the MDE context and they are easily distinguishable from other MDD tools because they explicitly focus on UML design, with limited support to the various MDE scenarios. For instance, they are different from metamodeling tools, which focus on providing the facilities for designing new modeling languages.

UML tools allow the creation of UML models, the import and export of them in XMI format, and sometimes they provide (partial) code generation facilities. The market currently offers a variety of licenses for UML tools, spanning from open source to freeware to commercial licensing. For instance, a comprehensive list of UML tools, describing the main features of each, can be found on Wikipedia.[6]

6.4.5   CRITICISMS AND EVOLUTION OF UML

UML has been criticized for being too verbose, too cumbersome, incoherent, impossible to use in domain-specific scenarios, and many other things, as reported, among others, in the famous article "Death by UML Fever" by Bell in 2004 [9]. A large number of disputes on how good or bad UML is can be found online too.[7] While some of these criticisms are probably partially true, UML remains the reference design language for software engineers. Furthermore, the criticisms have been taken seriously into account by OMG, which is putting a lot of effort in cleansing and simplifying the UML specification.

6.5   UML EXTENSIBILITY: THE MIDDLE WAY BETWEEN GPL AND DSL

If the development raises peculiar modeling needs, one may think to move toward a domain-specific language. Instead of opting for a completely new DSL, the language designer can decide for an intermediate solution, i.e., extending the existing GPL to fit the specific needs. To exemplify this solution, we show the extensibility features of UML.

Indeed, UML provides a wide set of extension features: stereotypes, constraints, tagged values, and profiles. These features are provided by the *Profile diagram*, which explicitly focuses on language extensibility. It operates at the metamodel level to show stereotypes as classes and profiles as packages. The extension relationship indicates what metamodel element a given stereotype is extending. This section aims at giving an overview of these extensibility options.

---

[6]http://en.wikipedia.org/wiki/List_of_UML_tools
[7]See, for instance: http://modeling-languages.com/ivar-jacobson-reflecting-evolution-uml, http://modeling-languages.com/grady-boochs-comments-umlmda, http://modeling-languages.com/uml-30-not-yet-or-so-it-seems, http://modeling-languages.com/complexity-uml.

6.5.1   STEREOTYPES

Stereotypes are used to extend meta-classes by defining additional semantics to the concept represented by the meta-class. Notice that the same model element may be stereotyped in multiple different ways. Stereotypes can be defined by specifying the following properties:

- *Base metaclasses*, defining what element(s) is going to be extended.

- *Constraints*, defining the special rules and semantics that apply to this specialization. This is what characterizes the stereotype in terms of semantics.

- *Tagged values*, defining zero or more values that the stereotype may need to know for working properly.

- *Icon*, defining the visual appearance of the stereotyped elements in the diagrams.

One doubt that may arise is why one should not use standard sub-classing instead of stereotyping. Indeed, this is an option, but the effect is quite different. Subclassing is a core operator of UML and basically defines a special relation between two items in the same model. Stereotypes instead let the designer define new modeling concepts, which in turn will be used for creating models. Stereotypes are typically used when you want to define (i) additional semantic constraints that cannot be specified through standard M1-level modeling facilities, (ii) additional semantics which may have significance outside the scope of UML (e.g., metadata such as instructions to a code generator), or (iii) the need arises for defining a new modeling concept which has some specific behavior or properties and which will be reused often in the modeling process.

6.5.2   PREDICATES

Another possibility for varying the semantics of UML is to apply *restriction predicates* (e.g., OCL expressions, as we will see at the end of this chapter) that reduce semantic variation of a modeling element. Predicates can be attached to any meta-class or stereotype in UML and can consist of formal or informal expressions. To be coherent with the standardized meaning of the UML symbols, the expressions must not contradict the inherited base semantics of the elements.

6.5.3   TAGGED VALUES

Tagged values consist of a tag-value pair that can be attached as a typed property to a stereotype. They allow the designer to specify additional information that is useful or required to implement, transform, or execute the model, e.g., for defining project management data (for instance, one can define a tagged value "status = unit_tested" to declare that a component has gone through the unit testing phase in the development process).

6.5.4 UML PROFILING

UML profiles are packages of related and coherent extensibility elements defined with the techniques described above. Profiles typically capture domain-specific variations and usage patterns for the language. In a sense, profiles are domain-specific interpretations of UML, and therefore, can be seen as domain-specific languages defined by extending or restricting UML. A UML profile example for describing the basic concepts of Enterprise Java Beans (EJB) is shown in Figure 6.10. The profile is composed of a set of stereotypes and relations (one of which is required).

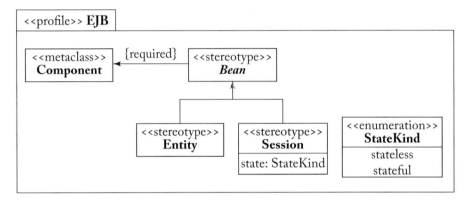

Figure 6.10: Example of UML profile for describing the basic concepts of Enterprise Java Beans (EJB).

While designers can define their own profiles, several UML profiles have been standardized and are widely recognized and used. Some examples of well-known profiles that became standards in OMG (we already mentioned them in Chapter 4 without highlighting they were UML profiles) are the following ones:

- UML Testing Profile (UTP): A UML profile that provides extensions to UML to support the design, visualization, specification, analysis, construction, and documentation of the artifacts involved in testing.

- OMG Systems Modeling Language (SysML).

- Service-oriented architecture Modeling Language (SoaML).

- UML Profile for System on a Chip (SoCP).

- UML Profile for Schedulability, Performance, and Time.

- UML Profile for Modeling and Analysis of Real-time and Embedded Systems (MARTE).

The complete list is available online on the OMG website.[8]

---

[8]http://www.omg.org/spec/#Profile

6.6   OVERVIEW ON DSLS

Domain-Specific Modeling Languages (DSMLs), also known as Domain-Specific Languages (DSLs), are design and/or development languages that are devised to address the needs of a specific application domain. DSLs are particularly useful because they are tailored to the requirements of the domain, both in terms of expressive power and notation.

The advantage is that a DSL does not force end users to study more general-purpose languages that may be full of concepts completely irrelevant for the domain, while at the same time providing appropriate modeling abstractions and primitives closer to the ones used in the domain.

6.6.1   PRINCIPLES OF DSLS

DSLs are a must-have for many application scenarios (not necessarily limited to software development). However, designing a DSL is not a trivial task: it implies good knowledge of the domain it applies to, as well as of language engineering practices. Finding the right abstractions for defining a DSL is hard and time consuming.

In order for the DSL to be useful, the following principles should be followed:

- The language must provide good abstractions to the developer, must be intuitive, and make life easier, not harder.

- The language must not depend on one-person expertise for its adoption and usage. Its definition must be shared among people and agreed upon after some evaluation.

- The language must evolve and must be kept updated based on the user and context needs, otherwise it is doomed to die.

- The language must come together with supporting tools and methods, because domain experts care about maximizing their productivity in working in their domain, but are not willing to spend a huge amount of time in defining methods and tools for that.

- A good DSL should be open for extensions and closed for modifications, according to the good old *open-close principle*, stating that "software entities (classes, modules, functions, etc.) should be open for extension, but closed for modification."

That's why good DSLs do not come out of the blue. The final goal of MDSE is not to make every developer a language designer. The goal is to have good language designers that build good DSLs for the right audience. Sometimes it takes years (together with tens of industrial experiences and thousands of users) to refine and polish a good language. This does not mean that nobody should develop a DSL, but one should be careful when designing one, because defining a new DSL is not a trivial task and may end up in a language which incorporates several weaknesses or even drawbacks for future users [44]. Accordingly, tools, design interfaces, and runtime architectures should be tailored to the domain as well, to avoid rejection of the language by the users.

Classification of DSLs

DSLs can be classified according to various dimensions, namely focus, style, notation, internality, and execution.

Focus

The focus of a DSL can be either vertical or horizontal. Vertical DSLs aim at a specific industry or field. Examples of vertical DSLs may include: configuration languages for home automation systems, modeling languages for biological experiments, analysis languages for financial applications, and so on. Horizontal DSLs have a broader applicability and their technical and broad nature allows for concepts that apply across a large group of applications. Examples of horizontal DSLs include SQL, Flex, IFML,[9] WebML,[10] and many others.

Style

The style of a DSL can be either declarative or imperative. Declarative DSLs adopt a specification paradigm that expresses the logic of a computation without describing its control flow. In other words, the language defines what the program should accomplish, rather than describing how to accomplish it. A typical example of declarative definition is the one of service choreography, which defines the one-to-one rules for Web service coupling. Imperative DSLs instead specifically require defining an executable algorithm that states the steps and control flow that needs to be followed to successfully complete a job. A typical example of imperative definition is the one of service orchestration, where a start-to-end flow of execution is defined between Web services.

Notation

The notation of a DSL can be either graphical or textual. The graphical DSLs imply that the outcomes of the development are visual models and the development primitives are graphical items such as blocks, arrows and edges, containers, symbols, and so on. The textual DSLs comprise several categories, including XML-based notations, structured text notations, textual configuration files, and so on.

Internality

As defined by Martin Fowler, external DSLs have their own custom syntax; you can write a full parser to process them, and you write self-standing, independent models/programs using them.

Internal DSLs instead consist in extending a host language to give the feel of a particular domain or objective to the host. This can be obtained either by embedding pieces of a DSL in the host language or by providing abstractions, structures, or functions upon it. Thanks to the embedding, you can reuse all the host language tooling and facilities.

---

[9]http://www.ifml.org
[10]http://www.webml.org

```
entity mux4_to_1 is
    port (I0,I1,I2,I3: in std_logic_vector(7 downto 0);
    SEL: in std_logic_vector (1 downto 0);
     OUT1: out std_logic_vector(7 downto 0);
end mux4_to_1;
```

Figure 6.11: Textual DSL example: VHDL specification of a 4 to 1 multiplexer.

Execution

As already discussed in Chapter 3, executability can be implemented through model interpretation (or code generation). Model interpretation consists in reading and executing the DSL script at run-time one statement at a time, exactly as programming languages interpreters do. Code generation instead consists of applying a complete model-to-text (M2T) transformation at deployment time, thus producing an executable application, as compilers do for programming languages.

6.6.2   SOME EXAMPLES OF DSLS

To make the discussion on DSLs more concrete, we report here a few excerpts of some of these languages. For the sake of generality, the exemplary languages are very diverse in terms of applicability and notation.

Figure 6.11 shows an example of textual syntax: the *VHDL language* (VHSIC Hardware Description Language) is a specification language for describing hardware electronic components. This small example shows how a multiplexer can be defined in VHDL.[11]

Figure 6.12a shows a requirement specification described with the goal-oriented notation *i** that describes the objectives of a health analysis system. Figure 6.12b shows a *feature model* describing the main aspects of a control system for a panel to be embedded in a wall.

Finally, Figure 6.13a shows a simple BPMN model (Business Process Model and Notation) which describes the process of hiring people in a company, while Figure 6.13b shows an excerpt of an IFML (Interaction Flow Modeling Language) model that describes the user interface of a website for navigating a product catalog by category.

6.7   DEFINING MODELING CONSTRAINTS (OCL)

Modeling and especially metamodeling languages are usually small, offering only a limited set of modeling constructs for defining new modeling languages.

This facilitates their use in practice (learning curve is not steep since language designers do not need to learn many different concepts) but it comes with a price. Metamodeling languages can only express a limited subset of all relevant information required in the definition of a modeling language. In particular, metamodeling languages only support the definition of very basic mod-

---

[11]A multiplexer (MUX) is an electronic component that selects one of several analog or digital input signals and forwards the selected input into a single line. See more at: http://en.wikipedia.org/wiki/Multiplexer.

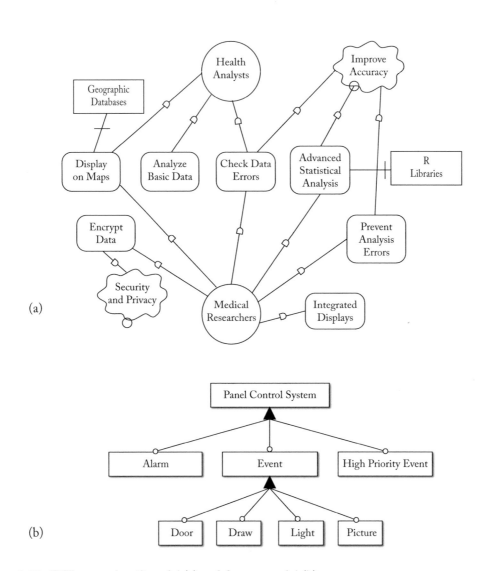

Figure 6.12: DSL examples: i* model (a) and feature model (b).

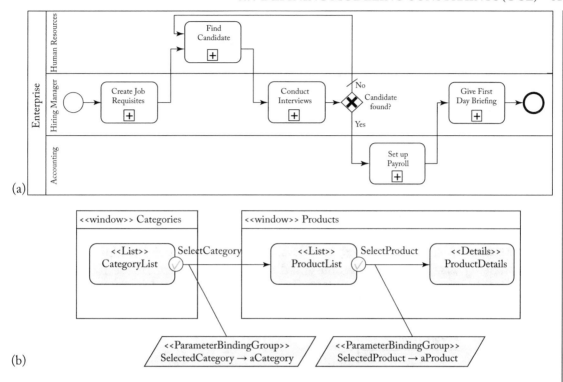

Figure 6.13: DSL examples: BPMN model (a) and IFML model (b).

eling constraints for the language-to-be, basically cardinality constraints that restrict the possible links between different modeling elements.

This is where Object Constraint Language (OCL) comes into play. OCL is a general-purpose (textual) formal language adopted as a standard by the OMG.[12] OCL is a typed, declarative and side effect–free specification language. *Typed* means that each OCL expression has a type, evaluates to a value of that type, and must conform to the rules and operations of that type. *Side effect-free* implies that OCL expressions can query or constraint the state of the system but not modify it. *Declarative* means that OCL does not include imperative constructs. And finally, *specification* refers to the fact that the language definition does not include any implementation details nor implementation guidelines.

OCL is used to complement metamodels with a set of textual rules with which each model conforming to that metamodel must comply.[13] When applied to a metamodel, these modeling

---

[12]OCL Specification. http://www.omg.org/spec/OCL
[13]As we will see throughout the book, OCL plays a key role in many other MDSE techniques, e.g., as part of the definition of matching patterns for model transformations.

constraints are also known as well-formedness rules since they define the set of well-formed models that can be specified with that specific modeling language.

Constraints in OCL are represented as invariants defined in the context of a specific type, named the *context type* of the constraint. Its body, the boolean condition to be checked, must be satisfied by all instances of the context type. The standard OCL library predefines the primitive and collection-related types (and their operations) that can be used in the definition of an OCL expression. Quantifiers (like *for all* and *exists*) and other iterators (*select*, *reject*, *closure*, etc.) are also part of the library. Access to the properties of an object and navigations from an object to its related set of objects (via the associations defined in the metamodel) is done using the dot notation. An introductory OCL tutorial can be found in `http://modeling-languages.com/object-constraint-language-ocl-a-definitive-guide`.

Some simple exemplary OCL expressions are reported below to let the reader appreciate the flavor of the language. Suppose we are describing a system for enterprise meeting planning, as represented in Figure 6.14, where the relevant classes are `Meeting` (comprising the attributes for describing `start` and `end` date/time and the `isConfirmed` attribute specifying whether the meeting is confirmed or not), `Team`, and `TeamMembers` (i.e., the people belonging to the team).

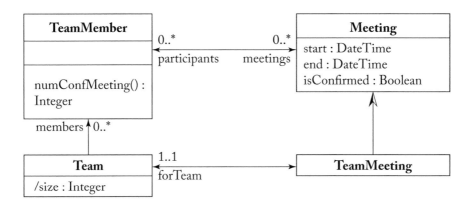

Figure 6.14: UML class diagram describing enterprise meetings (basis for the OCL examples).

The following invariant expression declares that for every `Meeting` instance the attribute end must be greater than the attribute `start`:

```
context Meeting
    inv: self.end > self.start
```

The next expression defines the derived attribute `size` in `Team` representing the number of members in the team:

```
context Team::size:Integer
    derive: self.members->size()
```

The next expression states that a meeting, if it is a team meeting, must be organized for all the members of the team:

```
context Meeting
    inv: self.oclIsTypeOf(TeamMeeting)
    implies self.participants->includesAll(self.forTeam.members)
```

Notice that the above rule uses `oclIsTypeOf` to check the type of the meeting. Another (more sensible) option would be to define the rule directly in the context of `TeamMeeting`.

Finally, the next example defines how the result of the method `numConfMeeting()`, determining the number of confirmed meetings of a `TeamMember`, is calculated. This is defined by using the post-condition operator (post).

```
context TeamMember::numConfMeeting():Integer
    post:
    result=meetings->select(isConfirmed)->size()
```

C H A P T E R   7

Developing your Own Modeling Language

Modeling has been often misunderstood as the process of just drawing pretty pictures. However, as we have already mentioned, models are much more than just pretty pictures. Models have to follow a clearly defined structure (exactly like program code), i.e., they have to conform to the associated *metamodel* representing the *abstract syntax* of the modeling language. Having a well-defined structure is the basis for applying operations on models such as loading/storing models from/to model repositories, querying and transforming the content of models, and checking the well-formedness of models to name just a few possible operations. In several modeling environments and tools, which use models as sketches or which were not built based on MDE practices, metamodels have a secret life hidden behind the user interface of the tools and, thus, are often invisible for the modelers. In simple usage scenarios, designers may not need to have a clear understanding of what is going on behind the curtains, but for complex MDE scenarios it is important to understand the big picture of MDE design. Especially, if you build your own tailored MDE environment, developing and using a metamodel is crucial, as we will see in this chapter. The main reason for this is that other language aspects going beyond the language's abstract syntax, such as the definition of the visual notation to be used when modeling, are heavily based on metamodels.

7.1   METAMODEL-CENTRIC LANGUAGE DESIGN

This chapter addresses the task of developing a new modeling language. In particular, we will delve into the task of defining the most important syntactical ingredients of modeling languages, namely the *abstract syntax* and the *concrete syntax* of the languages as introduced in Chapter 6.

The first ingredient is the definition of the modeling concepts and their properties by defining metamodels which play a role corresponding to grammars for textual languages.[1] While a grammar defines *all valid sentences* of a language,[2] a metamodel defines *all valid models* of a modeling language. Metamodels are defined with so-called *metamodeling languages*. Current metamodeling languages are heavily based on the core of structural object-oriented modeling languages, i.e., using *classes*, *attributes*, and *associations*—so to speak the core concepts of UML class diagrams. The term "metamodeling languages" originates from the fact that these kinds of modeling languages are applied for *modeling* modeling languages.

---

[1] In programming languages, EBNF-based grammars [35] are used for this purpose.
[2] For instance, the English grammar defines all valid English sentences; the Java grammar defines all valid Java programs.

The prefix *meta* in this context implies that the same technique is used to define the technique itself [47]. Another example for the usage of this prefix is *metalearning*, which stands for *learning how to learn*. Analogously, the term *metamodeling* stands for *modeling how to model*. The prefix *meta* may also be applied several times. Consider the term *meta-metamodeling*, which stands for modeling how to metamodel. The language for defining how to build metamodels is therefore called *meta-metamodel*—a model which is able to represent *all valid metamodels*. However, as models are always an abstraction, metamodels and meta-metamodels only define the abstract syntaxes of the languages they represent. Other concerns such as the concrete syntaxes or semantics are currently not covered by these kind of models and have to be specified with additional artifacts as we will see later.

Metamodels containing classes, attributes, and associations define the modeling concepts and their properties. However modeling constraints are only partially described (see also Chapter 6). For example, as known from UML class diagrams, multiplicity constraints may be defined for attributes and association ends as well as types for attributes. Additional constraints stating more complex validation rules for models may be defined based on these metamodel elements using a *constraint language*. As presented before in Chapter 6 for UML class diagrams, OCL is the language of choice for defining constraints going beyond simple multiplicity and type constraints. Thus, OCL is also reused for metamodels to define these kinds of constraints. Recall, metamodeling languages are heavily based on UML class diagrams, thus OCL can be also employed for metamodels. For instance, a modeling constraint such as "*A model element has to have a unique name*" is not expressible in current metamodeling languages which are based solely on class diagrams. However, this constraint is easily specifiable in OCL as we will see later in this chapter.

Metamodeling frameworks allow the specification of metamodels using dedicated editors as well as generating modeling editors out of the metamodels for *defining* and *validating* models. This means that metamodels are employed: (i) *constructively* by interpreting the metamodel as a *set of production rules* for building models and (ii) *analytically* by interpreting the metamodel as a *set of constraints* a model has to fulfill to conform to its metamodel.

All language aspects going beyond the abstract syntax of a modeling language have in common that they are defined on the basis of the metamodel as it is illustrated in Figure 7.1. Thus, in the rest of this chapter, we first show how to define the abstract syntax of a modeling language, including modeling constraints, and subsequently how graphical and textual concrete syntaxes on the basis of the abstract syntax are defined.[3] Furthermore, we elaborate on current technologies available on top of Eclipse which may be employed for developing these syntactical aspects of a modeling language.

Please note that the transformation aspects such as model-to-model transformations and model-to-text transformations are also based on the metamodels of modeling languages. How-

---

[3]Please note that graphical languages often comprise textual elements, e.g., in UML state machines the names of the states are rendered as text. Furthermore, there are also hybrid approaches for rendering the abstract syntax using textual elements such as supported by MPS, see: http://www.jetbrains.com/mps.

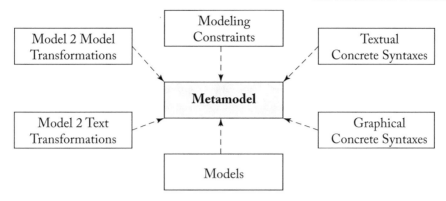

Figure 7.1: Metamodel-centric language design.

ever, before we elaborate on *transformation engineering* aspects, we first introduce the prerequisites in this chapter, i.e., *language engineering*. Transformation engineering is the subject of the two subsequent chapters (cf. Chapter 8 for model-to-model transformations and Chapter 9 for model-to-text transformations). By using transformations, the semantics of a modeling language can be formalized by following different approaches [45] as already introduced in Chapter 6: (i) giving *denotational semantics* by defining a mapping from the modeling language to a formal language; (ii) giving *operational semantics* by defining a model simulator; or (iii) giving *translational semantics* by defining, e.g., a code generator for producing executable code. While the first two approaches are using model-to-model transformations, the latter is often implemented as model-to-text transformations.

7.2   EXAMPLE DSML: SWML

In the following sections and chapters, we make use of a running example for demonstrating by example how modeling languages and model transformations are actually developed.

As context for the running example, we assume the following situation. A software development company is repeatedly building simple Web applications, which all comprise similar functionality. To be more specific, the Web applications are mostly used to populate and manage persistent data in a database. The Web applications are realized by using a typical three-layered architecture following the Model-View-Controller (MVC) pattern [27]. As implementation technologies, a relational database for persisting the data as well as plain Java classes for retrieving and modifying the data are employed for building the *model* layer. Apache Tomcat is used as the Web Server, and the *view* layer, i.e., the user interface, is implemented as *Java Server Pages* and the *controller* layer is realized as *Java Servlets*.

The functionality supported by the Web applications is the following. For each table defined in the database, the following Web pages are needed: (i) showing a list of all entries as an overview; (ii) showing the details of one selected entry; (iii) creating a new entry; (iv) updating an existing entry; and finally (v) deleting an existing entry. Although the functionality is more or less the same for all tables, it has to be re-implemented again and again for each table.

Space for improvements. To be competitive, the company has to develop high-quality Web applications within short time frames to have a unique selling point. In order to achieve this, the following goals have been identified by the company to further rationalize its software development process.

- **Homogeneity**: Web pages should provide the same interface and the same functionality for all tables.

- **Portability**: The transition to new devices such as mobile phones as well as to emerging Web development frameworks should be easily achievable.

- **Agility**: Early customer feedback is appreciated to allow for an agile development process.

- **Efficiency**: Repetitive tasks in the development process should be eliminated.

MDE to the Rescue. For realizing these goals, the company decided to switch from a traditional development approach to a model-driven approach. Because UML is designed as a general-purpose modeling language for object-oriented systems, it lacks specific modeling support for Web applications. Thus, the company decided to develop its own DSML called *simple Web Modeling Language* (sWML) for defining their specific kind of Web applications in a platform-independent way. Furthermore, platform-specific models following the MVC pattern should be derived with model transformations from which the Java-based implementations are finally generated. By following this approach, the company expects to gain the following benefits.

- **Homogeneity**: Model transformations always generate the same kind of implementation for a given type of model.

- **Portability**: Having platform independent models of the Web applications allows the support of new devices and frameworks by providing additional transformations.

- **Agility**: Prototypes for black-box testing are rapidly generated from models.

- **Efficiency**: Having a concise DSML and transformations eliminates repetitive tasks.

In the following sections and chapters, it is shown how this DSML is implemented based on Eclipse-based technologies. Of course, it is not the goal to present complete descriptions of the involved modeling languages and transformations. Instead, some interesting excerpts are shown to demonstrate the different concepts, techniques, and tools. The implementation artifacts of the DSML are provided on the book's website.[4]

---

[4]http://www.mdse-book.com

7.3 ABSTRACT SYNTAX DEVELOPMENT

In this section, we explain how modeling languages are developed using metamodeling as the central technique. The standard metamodeling language defined by the OMG is the *Meta Object Facility* (MOF) [56]. Besides this OMG standard, several languages and tools for metamodeling have been proposed in the last decade, the most prominent of which is the *Eclipse Modeling Framework* (EMF) offering the metamodeling language *Ecore*. MOF and Ecore are both based on a subset of UML class diagrams for describing structural aspects, whereas Ecore is tailored to Java for implementation purposes. Thus, for explaining how to define metamodels, we first use simplified UML class diagrams corresponding to MOF and afterward we discuss peculiarities of Ecore in Section 7.3.2.

One may raise the question: Why not just reuse UML for metamodeling as it is? Although MOF is very similar to the core of UML class diagrams, it is at the same time much more focused on the task of defining other modeling languages. For instance, MOF resigns n-ary associations, association classes, interfaces, and dependencies which are all concepts available in UML. The main differences between MOF and UML result from their fields of application. The domain of UML is object-oriented modeling in general. Thus, it is a comprehensive modeling language covering structural and behavioral modeling aspects as well as conceptual and implementation-oriented modeling. In contrast, the domain of MOF is much more focused—it is "simply" metamodeling. In a nutshell, MOF is a specialized DSML for metamodeling which reuses a subset of the UML core.

Benefits of metamodeling.   By using a metamodeling language, the language designer may define the abstract syntax of his/her own modeling language being either a general or a domain-specific modeling language. Independent of the kind of language which is metamodeled, having an explicit metamodel conforming to a standardized meta-metamodel comes with several benefits.

- **Precise language definition**: There is a formal definition of the language's syntax which is processable by machines. For example, the metamodels may be used to check if models are valid instances.

- **Accessible language definition**: With the core of UML class diagrams, a well-known modeling language is used to define metamodels. Thus, the knowledge of UML class diagrams is sufficient to read and understand the modeling language definitions in the form of metamodels.

- **Evolvable language definition**: Metamodels may be subject to modifications as any other model is. For example, the language may be extended with new modeling concepts by providing new subclasses for already existing metamodel classes. Having an accessible language definition further contributes to an easy adaptation of modeling languages based on meta-models.

In addition, generic tools, which are metamodel agnostic, may be developed based on the meta-metamodel level. For example, current metamodeling frameworks provide sophisticated reflection techniques to develop programs which are applicable to all instances of metamodels conforming to a particular meta-metamodel. Here are some examples.

- **Exchange formats**: Based on the meta-metamodel, there is support to serialize/deserialize models into XML documents which are exchangeable between tools supporting the same meta-metamodel.

- **Model repositories**: Similar to model exchange, models may be stored to and retrieved from a model repository based on generic storage/loading functionalities.

- **Model editors**: For modifying models, generic editors may be provided which are applicable on all models, irrespective of which modeling language is used.

The list of such generic support is not limited. For example, simple programs may be developed for computing model metrics, e.g., how many model elements are comprised by a model. But also sophisticated tool support may be developed based on the meta-metamodel level, e.g., model comparison and model versioning support as we will see later in Chapter 10.

Four-layer metamodeling stack.   Current metamodeling tools mostly employ a four-layer meta-modeling stack, as already introduced in Chapter 2. An overview on this metamodeling architecture is shown in Figure 7.2. The upper half of this figure is concerned with language engineering, i.e., building models for defining modeling languages, while the lower half is concerned with domain engineering, i.e., building models for particular domains. On the top layer—named M3 in OMG terms—there reside the metamodeling languages which specify the metamodeling concepts used to define metamodels. Normally, metamodeling languages are rather focused languages which offer a minimal set of concepts. In order to have a finite metamodeling architecture, these languages are most often reflexively defined, meaning that they are able to describe themselves. Thus, no additional language on top of M3 is needed to define M3, but you can think of having again the same language on M4 as on M3. Also, from a practical point of view, having a different language on M4 than on M3 seems to bring no further benefits. It would be again only a structural modeling language offering similar expressibility. MOF and Ecore are designed as reflexive languages typically used on the M3 layer.

On the M2 layer the metamodels reside which represent modeling languages (e.g., UML and ER) by defining their modeling concepts. These metamodels can be instantiated to build models on the M1 layer of the metamodeling stack. Models on the M1 layer represent systems such as a university management system (cf. UniSystem in Figure 7.2) and typically define domain

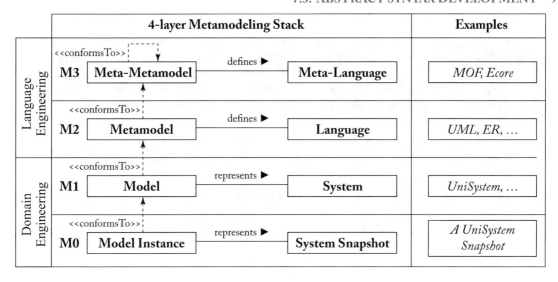

Figure 7.2: Four-layered metamodeling stack at a glance (based on [47]).

concepts by using the language concepts defined by the metamodels on M2. On the M0 layer there are the instances of the domain concepts which represent real-world entities, e.g., a *snapshot* of the university management system, at a given point in time.

The four-layer metamodeling stack assumes that a model on layer *M* conforms to a model on layer *M+1*. This relationship is called *conformsTo*. Thinking in graph structures, this means that a model is a graph consisting of nodes and edges. This graph has to fulfill the constraints of the next higher level. If each element (node or edge) on layer *M* is fulfilling the constraints given by its type element on layer *M+1* from which it is instantiated, the model on layer *M* conforms to its *type model*, i.e., the assigned model on layer *M+1*. In case of M0 and M1, the instances of a domain model have to conform to the domain model. In case of M1 and M2, the model has to conform to its metamodel. In case of M2 and M3, the metamodel has to conform to its meta-metamodel. Finally, in case of M3, the meta-metamodel has to conform to itself, because the *conformsTo* relationship is reflexive.

7.3.1   METAMODEL DEVELOPMENT PROCESS

One way to think about a metamodel is seeing it as the schema to store its models in a repository. Therefore, developing a metamodel is similar to developing a class diagram for a concrete domain. One has to identify the concepts that should be supported by the language, and, subsequently, these concepts have to be concisely specified by using a metamodeling language. As soon as the first metamodel version is available, the modeling language may be tested by using generic editors which allow the interpretation of the metamodel to build an example model. The editors may be used to model some reference examples to get a first impression whether all concepts have been defined properly or if some changes in the metamodel are necessary. To summarize, the metamodeling process in its simplest form is a three-step, iterative, and incremental process.

- **Step 1: Modeling domain analysis:** According to [43], three aspects have to be considered in the first phase of developing a modeling language: the *purpose*, *realization*, and *content* of the language. Especially, the last point is the most challenging: the identification of modeling concepts and their properties. For this purpose, the modeling domain that should be supported by the modeling language has to be analyzed. A pragmatic way to do this is to find several reference examples [62] that should be expressible in the modeling language to be developed—so to say the requirements for the language are defined by example.

- **Step 2: Modeling language design:** A metamodeling language is used to formalize the identified modeling concepts by modeling the abstract syntax of the language and modeling constraints should be formalized by using OCL. The output of this step is a metamodel for the modeling language to be developed.

- **Step 3: Modeling language validation:** The metamodel is instantiated by modeling the reference examples to validate the completeness and correctness of the developed metamodel. Furthermore, other general principles of language design such as simplicity, consistency, scalability, and readability also have to be considered. The result of this step provides important feedback for the next iteration of the metamodel development process.

Please note that this process is currently only concerned with the development of the abstract syntax of the language. To get feedback from domain experts, a concrete syntax is also needed in order to have an accessible modeling language. Thus, in practice, the complete process should also involve the development of the concrete syntax of the language. However, before we discuss how to develop a concrete syntax, the basics of developing the abstract syntax have to be understood. Thus, we discuss the development of the abstract and concrete syntax sequentially in this book due to didactic reasons, but we strongly encourage developing both syntactical concerns together in practical settings.

To make the metamodel development process more concrete, it is now instantiated for developing a metamodel for sWML.

Step 1: Modeling Domain Analysis

Several sources of information may be exploited for finding the appropriate set of modeling concepts. Which sources to tap depends on the purpose of the modeling language that should be developed. If a modeling language should be defined for abstracting from low-level program code, a major source of information may be existing programs. For example, recurring patterns found in the program code may be abstracted to modeling concepts. If the modeling language is tailored to a specific domain, document analysis or expert interviews may be a proper choice to derive the necessary modeling concepts. All these activities should lead to concrete reference examples which are beneficial for communication with domain experts, testing the modeling language and code generators, etc. Thus, we show how the reference example approach is used to develop the modeling concepts of the sWML language in the following.

Let us assume that a first sketch of a Web application model (cf. Figure 7.3) has been created during a workshop, based on an already existing Web application of a conference management system. For instance, the Web application model shows which talks and tutorials are offered by the conference. Based on this model sketch, we elaborate on the purpose, realization, and content of sWML.

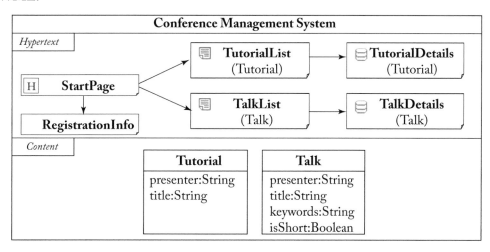

Figure 7.3: sWML model sketch.

Language purpose.   sWML should allow the modeling of the *content* layer and the *hypertext* layer of Web applications. The content layer defines the schema of the persistent data which is presented to the user by the hypertext layer in form of Web pages. In addition, the hypertext layer defines the navigation between Web pages and the interaction with the content layer, e.g., querying content from the database.

Language realization.   To support different sWML user types, a *graphical syntax* should be defined which may be used for discussing with domain experts and for reasoning on the structure of the hypertext layer and the content layer. However, for making a smooth transition from the standard development process based on programming languages to the model-driven approach, additionally, a *textual syntax* should be provided for developers who are familiar and used to working with text-based languages.

Language content.   A sWML model consists of a *content layer* and a *hypertext layer* which reflects the previously mentioned purpose of the language.

A content model contains an unrestricted number of *classes*. Classes have a unique name (e.g., Tutorial) and multiple *attributes* (e.g., Tutorial.title). Attributes have an assigned name and type. For instance, permitted types are: *String*, *Integer*, *Float*, *Boolean*, and *Email*. For each class, one attribute has to be selected as the representative attribute which is not explicitly indicated in the concrete syntax to keep the syntax concise.

Hypertext models contain different kinds of *pages*, whereas each page has a name. Exactly one page is the *homepage* of the Web application. Later on, this page (and all directly linked pages of the homepage) can be accessed by all other pages of the Web application. Pages are subdivided into *static* and *dynamic* pages. Static pages represent only *static content*, e.g., a collection of useful links. In contrast, dynamic pages represent *dynamically generated content* coming from the database. Thus, a dynamic page always has a relationship to a class defining the type of the displayed instances. The relationship to the used classes is shown in parentheses under the name of the page. Dynamic pages are further subdivided into *details pages* and *index pages* having specific icons. Index pages show all instances of a class in terms of a list, e.g., a list consisting of all tutorials of the conference, showing only the representative attribute of the instances. In contrast, details pages always show exactly one instance, e.g., one tutorial, with all its attributes.

Links represent the navigation between pages by pointing from a source page to a target page. On the source page, the link is visualized to the user. Thus, a page usually knows its links pointing to other pages but it is unaware of incoming links. Two kinds of links are distinguished: (i) non-contextual links (NCLinks) are standard links, which do not transport any information, e.g., a link to a static page; and (ii) contextual links (CLinks) transport information to the target page, e.g., to transfer data needed to compute the content of the target page, e.g., to select the instance shown by a details page.

Essence of the abstract syntax.   The first version of the language content description comprises different kinds of information. Primarily, the modeling concepts are introduced by defining their properties which is a necessary input for developing the abstract syntax. But sometimes other aspects such as the notation and semantics of the modeling concepts are discussed. Thus, from this description, the essence of the abstract syntax has to be filtered out, before the first version of the metamodel can be built. As a result, a list of concepts and their properties is produced. For our example, the table shown in Figure 7.4 summarizes the most important sWML concepts and their properties. In particular, *intrinsic* properties, having only primitive data values, must be distinguished from *extrinsic* properties, which represent relationships between modeling concepts.

Step 2: Modeling Language Design
The modeling concept table shown in Figure 7.4 guides the design of the metamodel as illustrated in Figure 7.5 for sWML. First, concepts are transformed to classes, intrinsic properties into attributes, and extrinsic properties are defined as associations between classes. Having this initial

Concept	Intrinsic Properties	Extrinsic Properties
Web Model	name : String	One *Content Layer* One *Hypertext Layer*
Content Layer		Arbitrary number of *Classes*
Class	name : String	Arbitrary number of *Attributes* One representative *Attribute*
Attribute	name : String type : [String|Integer|Float|…]	
Hypertext Layer		Arbitrary number of *Pages* One *Page* defined as homepage
Static Page	name : String	Arbitrary number of *NCLinks*
Index Page	name : String size : Integer	Arbitrary number of *NCLinks* and *CLinks* One displayed *Class*
Details Page	name : String	Arbitrary number of*NCLinks* and *CLinks* One displayed *Class*
NC Link		One target *Page*
C Link		One target *Page*

Figure 7.4: Modeling concept table for sWML.

structure, one can reason about further modeling constraints. For attributes, types have to be introduced such as String, Integer, and Boolean. If there is a range of possible values, enumerations may be defined as for the SWMLTypes (cf. Figure 7.5). For the association ends, upper and lower bounds of multiplicities have to be set properly.

Furthermore, one has to reason about the containment structure of the metamodel, because elements in models are often nested. For example, consider the content layer which contains classes, whereas classes again contain attributes. This containment structure has to be defined in the metamodel by declaring some association ends as compositions. The most important consequence of defining association ends as compositions is that when the container element is deleted all contained elements are deleted, too.

To enhance readability and extensibility of the language definition, inheritance relationships between classes are introduced to reuse attributes and associations (cf. abstract classes for pages and links). For this, refactorings for improving the structure of class diagrams may be applied on metamodels as well. For example, shifting up attributes from subclasses to superclasses, extracting superclasses from existing classes, or substituting an association with an explicit class to define additional properties for the association are recurring operations on metamodels. At the end of the day, metamodels should fulfill the well-known quality attributes of object-oriented modeling.

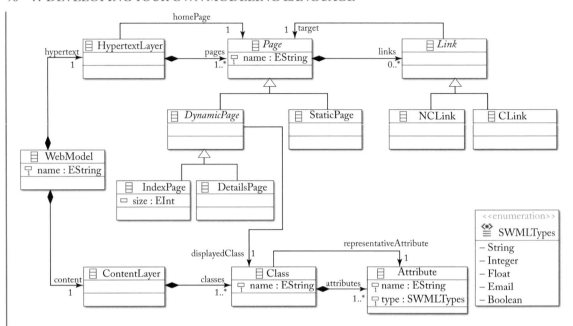

Figure 7.5: Metamodel for sWML.

Discourse.   For defining a modeling concept, we have the following three possibilities in meta-modeling languages: define it as a *class*, *attribute*, or *association*. Consider for instance the *homepage* concept in our example language. The first option (using a class) would result in an explicit class *HomePage*. However, if we would like to switch a homepage into a normal page, we have to delete the homepage, create a new page, and set all features using the previous values of the homepage for this new page. The second option (using an attribute) would result in having a Boolean at-tribute in the class *Page*, e.g., called *isHomepage*. Thus, for each page we can dynamically decide if the page represents a homepage or not. Of course, we may need a modeling constraint which ensures that for each Web application, only one homepage exists. Using the third option (an as-sociation), as we have done in the metamodel shown in Figure 7.5, allows us to mark exactly one page as the homepage within the set of all pages contained by the hypertext layer. Thus, we can dy-namically change the homepage and do not need to add an additional well-formedness rule. This discussion shows that when deciding how to represent a modeling concept in the metamodel, one has to reason about the advantages and disadvantages of using classes, attributes, or associations to represent the concept. This decision has a major impact on the actual modeling possibilities influencing the modeling experience the users will have. Thus, it is important to consider feed-back from users to improve the modeling language by switching between different metamodeling patterns.

Note that if a more powerful content modeling language is required, one option is to import existing metamodels as, e.g., the class diagram part of the UML metamodel, instead of remodeling the complete content modeling language from scratch.

Modeling constraints.   As already mentioned, several constraints cannot be defined by current metamodeling languages only in terms of graphical elements. Thus, OCL is employed to define additional constraints as so-called well-formedness rules. These rules are implemented in OCL as additional invariants for the metamodel classes and have to hold for every model. Thus, the constraints are defined on the metamodel and validated on the model level.

By introducing OCL invariants for metamodel classes, a modeling language is more precisely defined, leading to models with higher quality. This is especially important when employing models for code generation. Otherwise, some problems may not be detected until the final implementation is generated where the problems may manifest as compile-time and runtime errors, or remain undetected in the worst case.

In the following, we introduce some examples for well-formedness rules for sWML in natural language and subsequently show the corresponding OCL invariants.

Rule 1: A *Class* must have a unique name within the *Content Layer* to avoid name clashes.

```
context ContentLayer inv:
 self.classes -> forAll(x,y | x <> y implies x.name <> y.name)
```

Rule 2: The representative *Attribute* of a *Class* must be taken out of the set of *Attributes* contained by the *Class*.

```
context Class inv:
 self.attributes -> includes(self.representativeAttribute)
```

Rule 3: A page must not contain a non-contextual link to itself, because navigating such a link would result in exactly the same page as shown before the navigation.

```
context Page inv:  not self.links -> select(l | l.oclIsTypeOf(NCLink))
    -> exists (l|l.target = self)
```

Besides well-formedness rules, modeling guidelines and best practices can be defined with OCL invariants. For instance, naming conventions may be defined: the following rule imposes that all attribute names have to start with a lower case letter.

```
context Attribute inv:  if self.name.size() >= 1
 then     self.name.substring(1,1).toLower() = self.name.substring(1,1)
 else     false
 endif
```

This last example requires that a class should not have more than six attributes:

```
context Class inv:  self.attributes -> size() < 7
```

Step 3: Modeling Language Validation
Early validation of the abstract syntax can be achieved simply by manual instantiantion of the metamodel. In this case, we can immediately assess whether the designed metamodel is appropriately covering the modeling domain or not. Indeed, since we are using object-oriented meta-modeling languages like MOF to define metamodels, models consist of a collection of objects. Thus, it is common to represent models by UML object diagrams (cf. also Chapter 6). In UML, object diagrams allow the instantiation of class diagrams by using:

- *objects* for instantiating *Classes*;

- *values* for instantiating *Attributes*; and

- *links* for instantiating *Associations*.

Object diagram notation.   The notation of object diagrams is summarized in Figure 7.6. In the upper half of this figure, a class diagram is shown which is instantiated in the lower half by using an object diagram. Objects are notated similarly to classes using a rectangle, with the difference that in the first compartment the identifier of the object is given in combination with its type (e.g., a01 : ClassA in Figure 7.6). Of course, the type has to correspond to the name of a class in the metamodel. Attribute values are defined in the second compartment in so-called attribute slots. Links are notated like associations between objects.

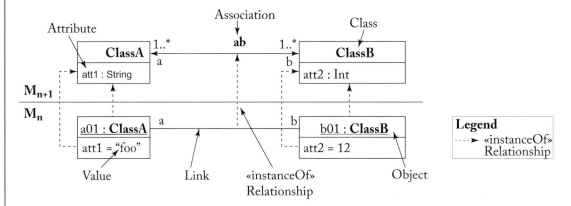

Figure 7.6: Object diagrams as instances of class diagrams.

For instantiating models from metamodels the same rules as instantiating object diagrams from class diagrams apply. First, only concrete classes can be instantiated. Second, data types such as enumerations act only as constraints for the attribute values, but cannot be instantiated. Third, the meta-features of attributes and references, i.e., multiplicities and uniqueness constraints, act only as constraints for the object diagrams. Finally, the containment references are also just represented as links in object diagrams. However, if a container object is deleted all directly and indirectly contained elements, i.e., objects linked by containment references, are automatically deleted. To enhance the readability, containment links are shown in the following by using the black diamond notation.

In Figure 7.7, the object diagram for an excerpt of the example model (cf. Figure 7.3) is shown. To better illustrate the relationships between (i) metamodel and model level and (ii) abstract and concrete syntax, the identifiers of the objects are annotated in the concrete syntax as special labels to the model elements. As can be seen in Figure 7.7, all model elements are represented as objects in the abstract syntax.

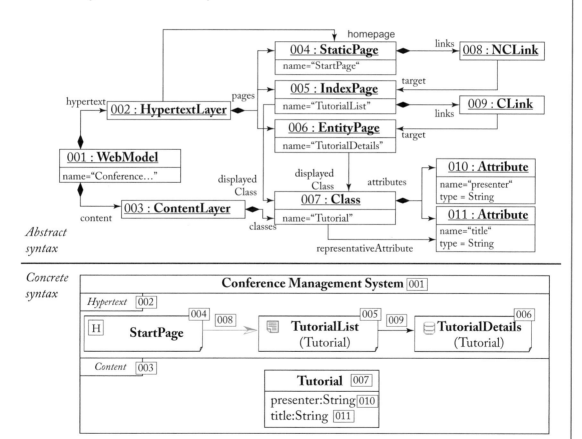

Figure 7.7: sWML model expressed using both the concrete and abstract syntax of the language.

Feedback for metamodel improvements.   When testing metamodels, several changes may be identified which are needed to represent and formalize the language properly. For example, the following simple modifications are quite common: mark classes as concrete/abstract, set references as containment/non-containment, restrict/enlarge multiplicities of features, make feature types more specialized/generalized, or just delete existing and introduce new elements. Not only simple modifications may be required, but also more complex changes such as refactorings for switching between metamodeling patterns, as explained before, or for the introduction of design patterns, e.g., the composition pattern, may be needed to improve the modeling language definition.

Note that having to change the metamodel in cases where instances of them already exist may lead to trouble when the changes break the conformance relationships between the metamodel and the models. For instance, the models may be no longer loadable in the modeling editors if metamodel elements are renamed or deleted. How to deal with such metamodel/model co-evolution problems is discussed in Chapter 10.

7.3.2   METAMODELING IN ECLIPSE

After having presented how metamodels are defined (using UML class diagrams) and instantiated (using UML object diagrams) on a conceptual level, we now discuss roughly how these tasks are supported by EMF (Eclipse Modeling Framework).[5] EMF comes with its own metamodeling language Ecore and tool support for defining Ecore-based metamodels as well as instantiating those. Please note that this subsection gives only a short overview on EMF and its core functionalities. For further information, we kindly refer the interested readers to dedicated EMF resources mentioned on our book website.[6]

An overview on Ecore's modeling concepts is depicted in Figure 7.8. The main concepts of Ecore are `EClassifiers` comprising `EClasses`, `EDataTypes`, and `EEnums`, as well as `EStructuralFeatures` comprising `EAttributes` and `EReferences`. Thus, the same object-oriented concepts are provided by Ecore as known from MOF. `EClasses` are the first-class citizens in Ecore-based metamodels. An `EClass` may have multiple `EReferences` and `EAttributes` for defining its structural features as well as multiple superclasses. An `EAttribute` is part of a particular `EClass` and has a lower- and an upper-bound multiplicity. Additionally, it can be specified as being ordered and unique in case the attribute is multi-valued. The type of an `EAttribute` is either a simple `EDataType` or an `EEnum` stating a restricted list of possible values, i.e., the `EEnum`'s literals (cf. `EEnumLiterals`). `EString`, `EBoolean`, `EInt`, and `EFloat` are part of Ecore's default data type set. Analogous to `EAttribute`, an `EReference` is part of a specific `EClass` and also has a lower and an upper bound multiplicity. An `EReference` refers to an `EClass` and optionally to an opposite `EReference` for expressing bi-directional associations. Besides, an `EReference` can be declared as being ordered, unique, and as a containment reference, i.e., if the container object

---

[5]`https://eclipse.org/modeling/emf`
[6]`http://www.mdse-book.com`

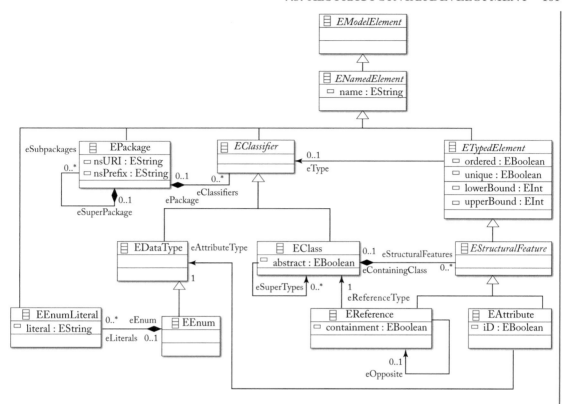

Figure 7.8: Overview of Ecore's main language concepts.

is deleted, all contained objects are deleted as well. Finally, `EPackages` group related `EClasses`, `EEnums`, `EDataTypes`, as well as other `EPackages`.

For specifying Ecore-based metamodels in Eclipse, there are several concrete syntaxes available supported by different editors. First, EMF comes with a tree-based editor for modeling metamodels in their abstract syntax similar to object diagrams, but using the containment hierarchy to form an explicit tree structure. However, as working with the modeling language's abstract syntax does not scale, it is recommended to use editors supporting concrete syntaxes for Ecore. There are several graphical editors for Ecore, e.g., the Ecore tools project,[7] which allow us to model Ecore metamodels using a similar syntax as UML class diagrams. In addition, there are other approaches which define Ecore metamodels textually such as KM3[8] and Emfatic.[9]

---

[7]http://www.eclipse.org/modeling/emft/?project=ecoretools
[8]http://wiki.eclipse.org/KM3
[9]http://wiki.eclipse.org/Emfatic

OCL support for Ecore.   For defining OCL constraints for Ecore-based metamodels several plugins are available in Eclipse. Eclipse OCL[10] is a project focusing—as the name suggests—on the implementation of the OCL standard within Eclipse. An interesting subcomponent in this project is OCLinEcore[11] which allows the definition of invariants for Ecore-based metamodels and the evaluation of these constraints within modeling editors. An alternative project is the Dresden OCL Toolkit.[12] Further support for defining and evaluating invariants for EMF-based models is provided by the Epsilon Validation Language[13] of the Epsilon project. The languages are inspired by OCL, but provide a slightly different syntactic appearance related to Java, as well as syntax extensions useful for practical aspects of model validation such as customizable error messages.

7.4   CONCRETE SYNTAX DEVELOPMENT

Modeling languages are often considered to be graphical languages, only. However, as there exist graphical programming languages, there is also the possibility to define a textual syntax for modeling languages. Thus, two kinds of concrete syntaxes are currently supported by existing frameworks: *Graphical Concrete Syntaxes* (GCS) and *Textual Concrete Syntaxes* (TCS).

As mentioned before, the abstract syntax development process is supported by several OMG standards. For the concrete syntax definition there is currently just one OMG standard available, namely the Diagram Definition (DD)[14] specification, which allows defining graphical concrete syntaxes. However, in the current UML metamodel definition in the UML standard document [57], the concrete syntax is only shown in so-called notation tables (one column is used for modeling concepts and one for their notation elements) and by giving some examples. A more formal definition of the UML concrete syntax is not given. This is in contrast to traditional text-based language engineering approaches where also the concrete syntax is formally defined by using, e.g., EBNF-based grammars.

Having the concrete syntax formally defined opens the door for using sophisticated techniques such as automatic generation of editors which allow us to manipulate the artifacts in their concrete syntax. Having such editors is a must for the productive usage of models in practice. Without having a formal definition of the concrete syntax, these editors have to be implemented manually, which requires much effort leading to high investments to realize a MDE environment.

Fortunately, there are several emerging frameworks which provide specific languages to describe the concrete syntax of a modeling language formally as well as generator components which allow the generation of editors for visualizing and manipulating models in their concrete syntax.

---

[10] http://www.eclipse.org/modeling/mdt/?project=ocl
[11] http://wiki.eclipse.org/MDT/OCLinEcore
[12] http://www.dresden-ocl.org
[13] http://www.eclipse.org/epsilon/doc/evl
[14] www.omg.org/spec/DD

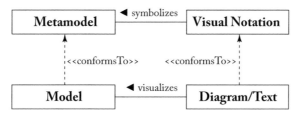

Figure 7.9: Visual notation: introducing symbols for modeling concepts.

Having a textual language allows the encoding of information using sequences of characters like in most programming languages, while graphical languages encode information using spatial arrangements of graphical (and textual) elements [51]. Thus, textual representations are one-dimensional, while most of the graphical languages allow for two-dimensional representations. For example, in UML diagrams each model element is located in a two-dimensional modeling canvas.

Regardless of the kind of concrete syntax developed, it is defined by mapping the modeling concepts described in the metamodel to their visual representations—so to speak, the visual notation introduces symbols for the modeling concepts (cf. Figure 7.9). The term "visual notation" is used in this chapter for summarizing both textual and graphical notations. Having the symbols for the modeling concepts allows us to visualize the models, either by using only text or also graphical elements arranged in a diagram.

In the following, we elaborate on different approaches for defining a visual notation for modeling languages and demonstrate how a GCS and TCS is developed for sWML.

7.4.1 GRAPHICAL CONCRETE SYNTAX (GCS)

Before we delve into details on how a GCS is defined for a metamodel, we first elaborate on the anatomy of a GCS.

Anatomy of graphical languages.   A GCS has to define the following elements:

- *graphical symbols*, e.g., lines, areas, complete figures such as SVG graphics, and labels for representing textual information, e.g., for visualizing the names of modeling elements;

- *compositional rules*, which define how these graphical symbols are nested and combined, e.g., a label visualizing the name of a model element is centered within a rectangle representing the model element;

- *mapping* of the graphical symbols to the elements of the abstract syntax for stating which graphical symbol should be used for which modeling concept, e.g., a specific model element type is visualized by a rectangle.

Current graphical modeling editors use modeling canvas, which allows the positioning of model elements in a two-dimensional raster. Each element has an assigned x,y coordinate, which normally stands for the upper-left corner of the graphical symbol. The model elements are mostly arranged as a graph which is contained in the modeling canvas. This graph is called diagram and represents a graphical view on the model. Please note that not all model information is actually shown in the modeling canvas. Several property values are only shown and editable in an additional property view. This, on the one hand, allows accessing and editing every property of a model element, while, on the other hand, avoids overloading the diagram with too much information.

In Figure 7.10, we show a small excerpt of a generic metamodel for GCSs. First, the metamodel states that a diagram consists of different elements which are either a node, edge, compartment, or label. These element types are sufficient for most modeling languages. For instance, nodes and edges are the important concepts to form *graphs* and are represented by shapes and lines, respectively. Compartments, mostly represented by shapes, are used to nest elements, i.e., diagrams are *nested graphs*. Finally, labels are used to annotate nodes and edges with additional information, i.e., diagrams are also *attributed graphs*.

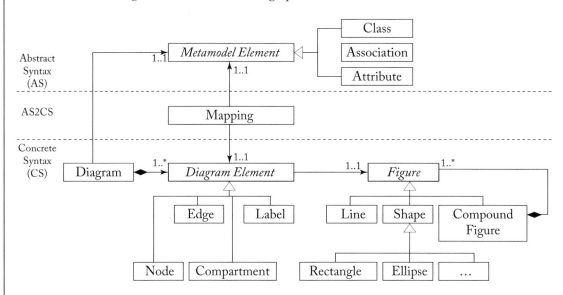

Figure 7.10: A generic GCS metamodel.

For stating the relationship between modeling concepts and symbols, we have a mapping between abstract syntax and concrete syntax elements. In general, an abstract syntax element, such as a class, attribute, and association, is mapped to a concrete syntax element.[15] For the diagram itself, the root element of the containment hierarchy of the metamodel is usually selected, which

---

[15]Mappings may be conditional, meaning that the selection of the element used for visualization depends on some feature values of the model elements.

is able to contain directly or indirectly all other model elements. All other metamodel classes are normally mapped to nodes and associations either to compartments in case of containment associations or to edges in case of non-containment associations. Attributes are mapped to labels residing within a node or are located near a node or an edge. There are also cases where a class may be mapped to an edge. Assume we have a class in the metamodel representing generalization relationships. In this case, instances of this class are represented as edges where the references of the class are used to determine the source and the target nodes of the edges.

If we take another look at Figure 7.7 (p. 99) which shows the abstract syntax for the reference modeling example given in sWML's concrete syntax, we may encounter the mapping between the modeling concepts and the symbols used in sWML. For example, we see immediately that Link objects are visualized as edges, while Page objects are visualized as nodes having specific icons attached based on their types.

Approaches to GCS development.   Before we demonstrate how to develop a GCS for sWML, we first give an overview on the three main GCS development approaches.

Mapping-centric GCS. Protagonists from this category provide dedicated modeling languages for describing the GCS and the mapping from the concrete syntax to the abstract syntax. This approach is followed by the Graphical Modeling Framework[16] (GMF). The language engineer has to define: (i) a .gmfgraph model which defines the graphical elements used to visualize model elements (cf. class Figure and DiagramElement in Figure 7.10); (ii) a .gmftool model which specifies the tool palette,[17] in particular, which icons are used to produce which model elements; and finally, (iii) a .gmfmap model which actually defines the mapping between elements defined in the metamodel and the graphical elements defined in the .gmfgraph model (cf. class Mapping in Figure 7.10). A generator transforms these three models into code, which represents the implementation of a fully fledged graphical modeling editor. Please note that GMF may be also seen as a DSML with a code generator component. Thus, the development of graphical modeling editors is achieved by applying MDE techniques.

Also, the OMG Diagram Definition (DD) uses this approach, exploiting MOF and QVT, and finally, also Sirius[18] uses mappings between abstract and concrete syntax elements.

Annotation-centric GCS. Approaches from this category directly annotate the metamodel with information about how the elements are visualized. This approach is supported by EuGENia.[19] The EuGENia framework allows us to annotate an Ecore-based metamodel with GCS information by providing a high-level textual DSML. This annotated information is used by a dedicated transformation component to generate the aforementioned GMF models. Therefore, EuGENia follows itself a model-driven approach to generate graphical editors by reusing GMF as the

---

[16]http://www.eclipse.org/gmf
[17]The tool palette is the area of a modeling tool offering icons for populating the modeling canvas by dropping elements into it.
[18]http://www.eclipse.org/sirius
[19]http://www.eclipse.org/gmt/epsilon/doc/eugenia

transformation target. EuGENia provides high-level annotations that hide the complexity of GMF and lowers the entrance barrier for creating initial versions of GMF-based editors. While EuGENia is very useful as a kick starter for developing graphical modeling editors, it doesn't stop there and can be also used all the way to the final polished version of the graphical editor.

API-centric GCS. There are also approaches allowing the implementation of graphical modeling editors directly on the code level by providing a dedicated programming framework.

This approach is taken by Graphiti.[20] Graphiti provides a powerful programming framework for building graphical modeling editors. A language engineer has to extend the provided base classes of Graphiti to define the concrete syntax of a modeling language. In particular, the framework provides classes for developing a *pictogram model* describing the visualization and the hierarchy of concrete syntax elements (similar to the .gmfgraph model) and a *link model* for establishing the mapping between abstract and concrete syntax elements (cf. .gmfmap model of GMF).

Another protagonist is the Graphical Editing Framework[21] (GEF), which provides low-level functionalities for graphical editors. Thus, GEF is considered to be the infrastructure for other GCS approaches rather than a framework for directly implementing graphical modeling editors. Although this would be possible, much more effort is required.

Finally, GMF also provides a powerful API for programming modeling editors, but it is recommended to start with the GMF models and only refine the generated editors on the code level. Thus, GMF is classified as a mapping-centric approach, although it is possible to start programming a modeling editor from scratch.

Defining a GCS for sWML in EuGENia.   To give the reader an idea how to define a GCS for a concrete example, we employ EuGENia for developing a GCS for sWML. We selected EuGE-Nia, because it allows us to introduce GCS approaches on an appropriate level of abstraction. In EuGENia there are several annotations available for specifying the GCS for a given Ecore-based metamodel. In the following, the main annotations are first enumerated and subsequently applied for sWML.

- **Diagram**: The root element of the abstract syntax representing the model, i.e., the element containing (directly or indirectly) all other elements, is a perfect match for representing the modeling canvas. EuGENia uses the `Diagram` annotation to mark this class in the metamodel.

- **Node**: Instances of metamodel classes are often visualized as nodes within the diagrams. Thus, EuGENia allows annotating classes with the `Node` annotation. This annotation has several features, such as selecting the attribute of the annotated class which should be used as the label for the node, layout information such as border styles, colors, and either an

---

[20]http://www.eclipse.org/graphiti
[21]http://www.eclipse.org/gef

external figure (e.g., provided as a SVG graphic) or a predefined figure by EuGENia (e.g., rectangle or ellipse) may be used to render the node.

- **Link**: The Link annotation is applicable to classes as well as to non-containment references that should appear in the diagram as edges. This annotation provides attributes for setting the style of the link, e.g., if it is dashed, and the decoration of the link end, e.g., if the link end should be visualized as an arrow. For classes annotated with the Link annotation, additionally, the source and target references used as end points for the link have to be selected from the available references of the classes.

- **Compartment**: Containment references may be marked with this annotation. It defines that the containment reference will create a compartment where model elements that conform to the type of the reference can be placed within.

- **Label**: Attributes may be annotated with the Label annotation which implies that these attributes are shown in addition to the label used as name for the node or link.

We now discuss two excerpts of the GCS definition for sWML (cf. Figs. 7.11 and 7.12). Annotating the metamodel, as in the upper part of the figures, allows us to render the models as schematically illustrated in the bottom of the figures.

The first excerpt (cf. Figure 7.11) comprises three annotations. The first one is used to select the element which is used to represent the diagram. Thus, the class WebModel (the root element of sWML) is annotated with Diagram. Furthermore, the diagram should allow placing a hypertext layer within the canvas. Thus, a Compartment annotation is necessary for the containment reference hypertext. This annotation states that it is possible to add a hypertext layer to the diagram. Finally, hypertext layers should be visualized as nodes represented as rectangles. Thus, we employ a Node annotation to the HypertextLayer class and set the figure attribute accordingly. If no label is specified, the default label for a modeling element is its type name.

The second excerpt (cf. Figure 7.12) defines that instances of the Page class are visualized as rectangles[22] where the name of the page is shown as the label of the rectangle. This is achieved by a similar Node annotation as before for the HypertextLayer class in Figure 7.11. Instances of the Link class should be visualized as edges between the nodes representing the pages. Thus, the class Link is annotated with a Link annotation. Several attributes have to be defined for the proper visualization, such as which references are used as the *target* for the links (the source for a link is per default its container which is applicable for our case) and how the links should actually look. In this example, the link end at the target side should be visualized as an arrow to indicate the navigation direction of the hyperlink. Please note that the GCS definitions defined for superclasses are applicable also for their subclasses in case no specific definitions have been defined (i.e., in Figure 7.12 also the graphical notation for contextual and non-contextual links is defined).

---

[22]For the sake of simplicity, we do not discuss the complete GCS specification for pages such as icons, etc.

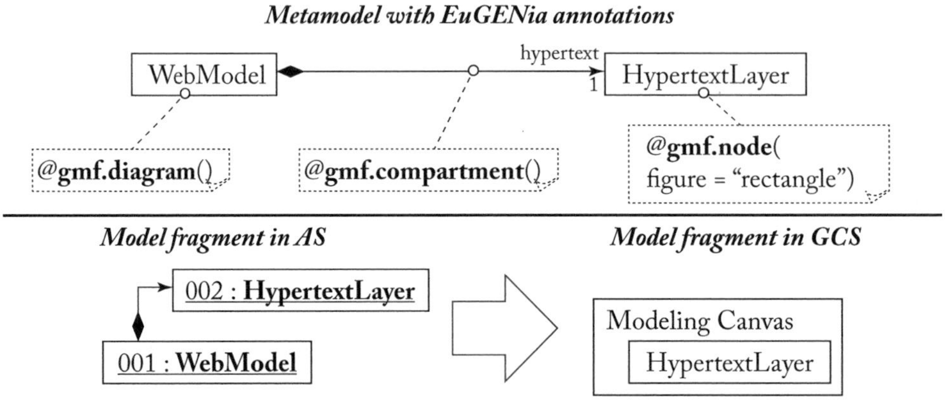

Metamodel with EuGENia annotations

Figure 7.11: GCS excerpt 1: Diagram, Compartment, and Node annotations.

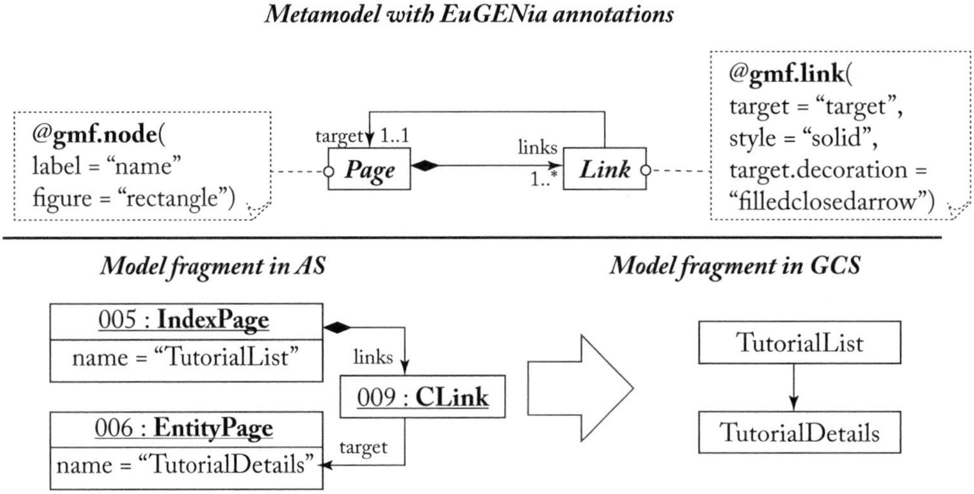

Metamodel with EuGENia annotations

Figure 7.12: GCS excerpt 2: Node and Link annotations.

7.4.2   TEXTUAL CONCRETE SYNTAX (TCS)

In the early days of MDE, modeling languages have been considered to be solely graphical languages mainly influenced by precursors of UML. However, in the meantime, this view has changed fundamentally. First, the usefulness of having a TCS for a modeling language has been recognized in domains where users are familiar working with textual documents. Second, powerful frameworks emerged which allow for the user-friendly development of textual editors for modeling languages.

The fundamental assumption of textual specifications is that the comprised text consists of a sequence of characters. Of course, for defining meaningful specifications, not every arbitrary sequence of characters represents a valid specification. Thus, a grammar is needed that specifies all valid character sequences. As we will see later, from a metamodel, only a generic grammar may be derived which allows the generic rendering of models textually as well as the parsing of text into models. However, syntactic sugar is highly appreciated when working with textual languages. In particular, language-specific keywords enhance the readability of textual specifications a lot. Thus, some additional concepts are needed for developing a TCS.

Anatomy of textual languages.   As we have seen before, a graphical syntax consists of several different types of elements such as different kinds of geometrical figures. Also textual languages make use of different types of elements which are explained based on our running example sWML. Let us assume we have an excerpt of the conference model as illustrated in the left-hand side of Figure 7.13. In the hypertext layer we have the IndexPage TutorialList which visualizes all instances of the Class Tutorial defined in the content layer. A possible textual visualization of this example is shown in the right-hand side of Figure 7.13 which provides the same information as the graphical visualization, but using a TCS.

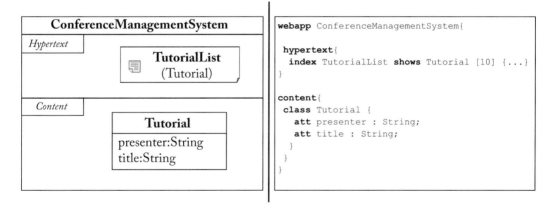

Figure 7.13: sWML model excerpt in GCS and TCS.

By taking a closer look at this example, the following kinds of TCS elements can be identified which are of paramount importance for every TCS, in general.

- **Model information:** Obviously, a TCS has to support the model information stored in the abstract syntax. In our example, we have to define the *name* of the model elements and for attributes also the type. This is analogous to graphical languages where labels are normally used for stating such information.

- **Keywords:** A keyword is a word that has a particular meaning in the language, i.e., it stands for a particular language construct. In our example, keywords are used for introducing the different model elements. The terms used for representing keywords are reserved words and cannot be used as values for model elements. For example, a class called *class* would not be possible without using special markers which eliminate the keyword character of the term.

- **Scope borders:** In a GCS, a figure defines the borders of a model element. In a TCS no figures are used but instead special symbols, so-called scope borders are used to mark the beginning and the end of a certain section. In our example, curly brackets well known from programming languages are used as scope borders. After introducing an element by its keyword, an opening curly bracket and a closing curly bracket are used to define the compartments of the element. Thus, the scope borders are also of special importance for nesting elements in a TCS.

- **Separation characters:** In a textual artifact a list of elements may be specified at a certain position. In such cases, a special character is needed for separating the entries of the list. In our example, we are using the semi-colon to separate the different attributes introduced for a class.

- **Links:** In a GCS, edges are used to link different elements related by non-containment references. In a textual language, we only have one dimension to define elements. Thus, links which are not defined as containments cannot be explicitly visualized. To specify links, identifiers have to be defined for elements which may be used to reference an element from another element by stating the identifier value—similar to the foreign key concept in relational databases. In our example, the page has to be linked to a class of the content layer. Thus, the class has to provide an identifier. A natural identifier of the class is its name, as it is known from programming languages to use class names as types.

Approaches to TCS development.   Besides the model information, metamodels do not provide information about the other kinds of TCS elements. For the definition of this TCS specific information, two approaches are currently available in MDE: (i) having either a generic TCS or (ii) a language-specific TCS.

Generic TCS. Such as for XML, a generic TCS may be also defined for models. This means, similar to using object diagrams to graphically visualize models in a generic way, a textual syntax generically applicable for all kinds of models may be applied. You can think of this textual syntax as a textual format for specifying object diagrams. The benefit is that the metamodel is sufficient to derive a TCS, i.e., no additional concrete syntax specification is needed. A drawback is that no tailored syntax can be developed dealing with the specifics of a given modeling language.

This approach has been used to define the XMI syntax for serializing models into XML documents as well as the Human Usable Textual Notation (HUTN), both standardized by the

OMG. An implementation of the HUTN standard is provided for EMF by the Epsilon HUTN project.[23] How models are generically serialized into XML documents is presented in Chapter 10.

Language-specific TCS. To eliminate the drawbacks of having solely a generic TCS, approaches for defining specific TCSs for modeling languages have been proposed. With respect to the methodology of the language engineering process, two approaches can be distinguished.

Metamodel first. To develop the syntax of a modeling language, a metamodel first approach may be followed. This means, first the abstract syntax is defined by the means of a metamodel. In a second step, the textual syntax is defined based on the metamodel. Metamodel first approaches are based on the assumption that metamodels represent the central language artifacts (cf. Figure 7.1, p. 87), thus the concrete syntax should be defined on top of the abstract syntax. This approach is also taken by the aforementioned GCS approaches. For defining a TCS, for each metamodel class, a text production rule may be defined for rendering the model elements into a text representation. The production rules consist of a left-hand side (stating the name of the rule) and right-hand side (describing valid terminal and non-terminal symbol sequences) as we will see later in this section by going through a concrete example.

This approach is followed by the Textual Concrete Syntax (TCS) project[24] which allows for the definition of a textual syntax based on text production rules similar to EBNF but with specific extensions for taking care of the specific nature of models such as their graph-based nature. TCS allows us to parse text into models and to pretty-print models as text. EMFText[25] is another protagonist of a metamodel first approach for EMF models.

Grammar first. This kind of approach follows the same goal as metamodel first approaches, but proposes a different methodology to reach this goal. Inspired by EBNF, these approaches start the language definition by developing the grammar defining the abstract and concrete syntax at once as a single specification. The languages to define the grammars are also based on text production rules as used for metamodel first approaches. In a subsequent step, the metamodel is automatically inferred from the grammar by dedicated metamodel derivation rules.

For instance, this approach is originally followed by Xtext[26] for developing TCS-based languages for EMF.[27] Monticore[28] also follows a grammar-first approach for developing textual DSLs within Eclipse, but this project is not based on EMF and uses its own model format.

At the end of the day, both language-specific TCS development methodologies result in the same artifacts: (i) having a metamodel for the abstract syntax and (ii) having a TCS for the models. Thus, it is mostly a matter of familiarity with which kind of approach you want to work. The

---

[23]http://www.eclipse.org/epsilon/doc/hutn
[24]http://www.eclipse.org/gmt/tcs
[25]http://www.emftext.org
[26]http://www.eclipse.org/Xtext
[27]Xtext has been extended to allow also for a metamodel first development approach by generating a default TCS from a metamodel.
[28]http://www.monticore.org

generic TCS approach is not suitable for producing a specific TCS, but has the advantage of having a low cost textual serialization of models. However, working with large models shown in HUTN is too complicated in a practical context. But it may be considered as a bootstrapping technology for deriving a first version of a grammar from a metamodel which is subsequently extended manually. For instance, EMFText provides such capabilities to derive a HUTN/Java based TCS automatically from an Ecore metamodel. Xtext also allows us to generate a default TCS from a metamodel to speed up the grammar development and to ease the synchronization between the metamodel and grammar.

Defining a TCS for sWML in Xtext.   We selected Xtext for demonstrating the development of a TCS, because of its mature tool support within Eclipse. Xtext provides a grammar definition language similar to EBNF, but with additional features to achieve a similar expressivity as meta-modeling languages such as Ecore. This is an important extension, because from the grammars, metamodels are automatically generated which should exploit all possibilities of metamodeling languages to produce high-quality metamodels. Having a metamodel, a grammar, model-to-text serializer, and text-to-model parser for a modeling language allows a smooth transition from text to models and vice versa. Thus, having a language defined with Xtext allows the use of all the tool support available for EMF-based models such as model transformations, code generation, etc.

Besides the interoperability between text-based and model-based representations, a text-based editor is automatically generated from the grammar definition. The generated editor supports syntax checking and highlighting, code completion, and well-defined extension points to further enhance the editing functionality programmatically using Java. Context-sensitive constraints are described in an OCL-like language called Check which are formulated against the automatically generated metamodel.

Text production rules comprise the core of Xtext-based grammars. In particular, three kinds of production rules are distinguished.

- **Type rules:** Type rules are the counterpart of classes in metamodels and are used to define modeling concepts. Consequently, when generating a metamodel from an Xtext-based grammar, a corresponding class in metamodels is produced, whereas the name of the rule corresponds to the name of the class. Type rules contain (i) *terminals* which represent the keywords, scope borders, and separation characters of the language and (ii) *non-terminals*. Non-terminals can be further distinguished into *assignments* which are mapped to attributes or containment references, and *cross references* which are mapped to non-containment references when a metamodel is generated from the grammar. So to speak, the non-terminals represent the features of a type rule. Thus, for each non-terminal, a feature is generated in the corresponding metamodel class. For defining assignments, several operators are available for setting the multiplicities of the features.

- **Terminal rules:** Such rules are similar to EBNF rules and are used to just return a value, i.e., a sequence of characters. In contrast to EBNF rules, terminal rules may have an assigned return type such as String or Integer.

- **Enum rules:** These rules are used for defining value enumerations. Thus, they are simply transformed to EEnums in the corresponding metamodels.

In the following, we demonstrate the use of *type rules*, *terminal rules*, and *enum rules* by defining an excerpt of the TCS definition for sWML (cf. Listing 7.1) which is sufficient for rendering the example model shown in Figure 7.13.

Listing 7.1: Xtext grammar excerpt of sWML

```
grammar book.SWML with org.eclipse.xtext.common.Terminals
generate sWML "http://book.SWML"
import "http://www.eclipse.org/emf/2002/Ecore" as ecore

WebModel :
  'webapp' name=ID '{'
     hypertext=HypertextLayer
     content=ContentLayer
  '}' ;

HypertextLayer :
  'hypertext {'
     pages+=IndexPage+
  '}' ;

IndexPage :
  'index' name=ID 'shows' displayedClass=[Class] '[' resultsPerPage ']' '{' ... '}' ;

terminal resultsPerPage returns ecore::EInt :    ('10' | '20' | '30') ;

ContentLayer :
  'content {'
     classes+=Class+
  '}' ;

Class :
  'class' name=ID '{' attributes+=Attribute+ '}' ;

Attribute :
  'att' name=ID ':' type=SWMLTypes ';'  ;

enum SWMLTypes :
  String | Integer | Float | Email | Boolean ;
```

When we compare the resulting grammar with the sWML metamodel shown in Figure 7.5, we see that the assignments for defining the *name* attributes of the modeling concepts are defined as IDs and not as Strings. This is important for allowing cross references in the textual specifications. For instance, such a cross reference is needed for the type rule *IndexPage* which defines a cross reference to *Class* (cf. displayedClass=[Class]). This assignment defines that after the keyword "shows," a value has to be defined which is equivalent to the name of a class. All other assignments representing references are defined as containment references, e.g., cf. hypertext=Hypertext. Furthermore, also one terminal rule is defined for the values used for configuring the maximum number of results shown per page. This means, after the reference to the presented class, the number of instances visualized per page is given in square brackets. Please note that this feature may also be modeled as a normal Integer attribute in combination with an additional OCL constraint for restricting the values to fit either 10, 20, or 30. The assignments representing attributes and cross references are defined as single-valued assignments (no additional operator for multiplicity is used), whereas the assignments representing containment references are all defined to be multi-valued (+ operator is used in addition to the = operator).

In Figure 7.14, the automatically derived metamodel for the Xtext grammar of Listing 7.1 is shown. Please note that the lower multiplicities of all features are automatically set to zero. Further, in the automatic derivation process, the ID attributes of the Xtext grammar are translated to String attributes in the metamodel which corresponds to the manually developed sWML metamodel. Finally, the terminal rules are not explicitly represented in the derived metamodel.

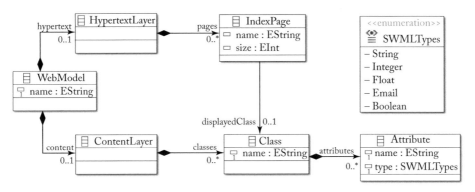

Figure 7.14: Automatically derived metamodel from the Xtext grammar.

7.5   A REAL-WORLD EXAMPLE: IFML

In order to provide a better understanding of what it means to define a modeling language in a realistic setting, going beyond the toy examples discussed so far, we provide here an example of a real-world modeling language definition. In particular, we discuss the Interaction Flow Modeling Language (IFML) that was adopted as a standard by the Object Management Group (OMG) in 2014.

IFML supports the specification of the front-end of applications, independently of the technological details of their realization. Thus, it can be seen as a richer (superset) modeling language in comparison to sWML, presented so far. With respect to sWML, IFML addresses any kind of software front-end modeling, covering: (i) the composition of the view, in terms of visualization units that compose the interface; (ii) the content of the view, i.e., what information is displayed to the user; (iii) the commands and interaction events supported by the user interface; (iv) the actions, i.e., the business logic that is triggered by the events; and (v) the parameter binding, defining the data items that are communicated between the elements of the user interface.

7.5.1 REQUIREMENTS

Four main requirements need to be addressed when defining a language that aims at becoming widely used and adopted:

- Definition of the language **based on standard modeling frameworks**;

- Definition of the language features by keeping into account the **usability and simplicity** of adoption, besides the **coverage of the domain** requirements;

- Capability of **integration with other modeling perspectives**, typically by allowing referencing to modeling elements or whole diagrams specified using other languages;

- **Extensibility of the language** through appropriate and limited extension points that allow us to cover specific domain or enterprise needs, without distorting the nature of the language; and

- Support for **reuse** of modeling elements, so as to allow for minimal redundancy in models.

7.5.2 FULFILLING THE REQUIREMENTS IN IFML

The above requirements have been addressed in the definition of the IFML language as follows:

- **Standardization:** The IFML definition is based on the OMG standards, i.e., OMG Model Driven Architecture (MDA) framework. In particular, the specification consists of five main technical artifacts. The *IFML metamodel* and its description specify the structure and basic semantics of the IFML constructs. The *IFML UML profile* defines a UML-based syntax for expressing IFML models. In particular, the IFML UML profile is based on the use of UML components (both basic components and packaging components), classes and other core UML concepts. The *IFML visual syntax* offers a concrete visual representation of the model. The *IFML serialization* provides a standard model interchange format for ensuring tool interoperability. All these parts are specified according to the OMG standards. The metamodel is defined through the MOF metamodeling language (an equivalent Ecore definition is available as well). The UML profile is defined according to UML 2.4 profiling rules. The visual syntax is defined through Diagram Definition (DD) and Diagram Interchange (DI) standards. The model serialization and exchange format is based on XMI.

- **Simplicity:** IFML expresses the user interaction modeling requirements using a visual modeling language, which has been validated based on the extensive usage of some of its variants over the years, exploiting the existing "ancestor" languages such as WebML [14]. This allowed us to clean and prune the language from unnecessary complexity. The design of IFML adheres as much as possible to the following "golden" rules: conciseness, i.e., minimum number of diagram types and concepts needed to express the salient interface and interaction design decisions; inference from the model, i.e., whenever something can be deduced from existing parts of the model, inference rules at the modeling level can apply default modeling patterns and details, avoiding the need for modelers to specify redundant information. One of the main simplification factors is that the IFML visual syntax offers a concrete representation based on a unique diagram type. This compacts all the aspects of the user interface that are otherwise expressed separately in other approaches, such as in MobML [26] or in UML (e.g., cf. the UML profile WAE [18]) which require a combination of class diagrams, state machines, sequence diagrams, activity diagrams, composite structure diagrams, and custom diagrams for describing the user interfaces. The reduction to a single-diagram representation contributes to making IFML a modeling language quickly learned, easy to use, and easy to implement by tool vendors.

- **Integration:** Sometimes the hardest choice in language design is what to leave out. This decision may arise from two main motivations: (i) there are aspects that are not of interest for the use of the language, and thus should not be considered; and (ii) there are aspects that are not the core contribution of the language, and thus, despite being related to the domain of interest of the language, are delegated to other existing languages that focus on those aspects. In the latter case, the best choice is indeed to cover the core aspects directly in the language, and then connect (i.e., reference) other languages for the remaining issues. For instance, IFML purposely ignores presentation aspects, because presentation is adversarial to abstraction (in graphic design every pixel is important) and does not specifically connect to any modeling language for that part. On the other side, it delegates to external models the specification of aspects that, although relevant to the user interface, are not parts of it, such as data persistence, synchronization of widgets, and definition of business logic.

- **Extensibility:** IFML builds upon a small set of core concepts that capture the essence of interaction: the interface (containers), stimuli (events), content (components and data binding), and dynamics (flows and actions). By design, these concepts are meant to be extended to mirror the evolution of technologies and devices. Thus, IFML incorporates standard means for defining new concepts as specializations of the ones defined in the core language. For instance, a designer can study new components types (representing specific widgets) or new event types for a particular domain or device.

- **Reuse:** IFML helps language users to define reusable and modularized models through the definition of dedicated model elements called *Modules*, specifically studied for supporting the definition of generic (and parametric) pieces of user interaction that can be referenced in any part of the design.

7.5.3 METAMODELING PRINCIPLES

The IFML metamodel is defined according to the best practices of language definition, including abstraction and modularization.

Abstraction. It is a good practice to define language concepts at different levels of abstraction, so as to separate more abstract and general concepts from low-level concrete concepts. Core concepts are extended by concrete concepts to cater to more precise behaviors. Multiple levels of abstraction can be devised, where subsequently more precise or detailed concepts inherit from the general ones.

Modularization. With the term modularization we mean the practice of splitting a language definition into coherent and cohesive modules. This is crucial for language understandability. The most typical way for implementing modularization in a language is by aggregating metaclasses into coherent packages. The IFML metamodel is divided into three packages: the *Core* package, the *Extension* package, and the *DataTypes* package.

The *Core* package contains the concepts that build up the infrastructure of the language in terms of InteractionFlowElements, InteractionFlows, and Parameters.

The *Extension* package contains the concrete concepts that specify the Core package concepts to cater to more precise behaviors, according to the abstraction principle.

The *DataTypes* package contains the custom data types defined by IFML. The IFML metamodel reuses the basic data types from the UML metamodel and specializes a number of UML metaclasses as the basis for IFML metaclasses. Furthermore, it assumes that a domain model is represented with a UML class diagram or with an equivalent notation.

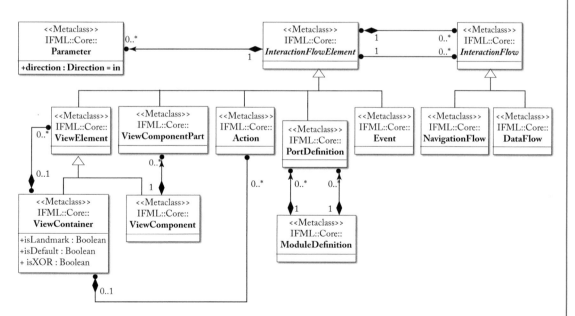

Figure 7.15: IFML Metamodel excerpt: InteractionFlow and InteractionFlowElement.

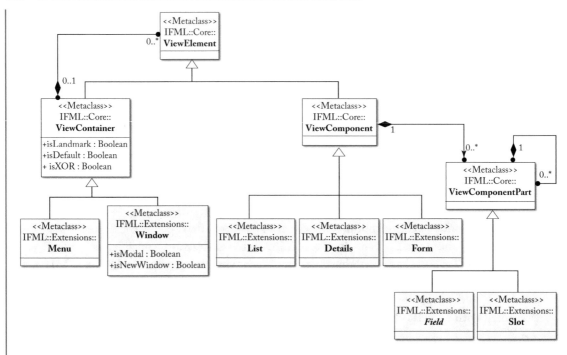

Figure 7.16: IFML Metamodel excerpt: ViewElement and its hierarchy (and containment).

7.5.4   IFML METAMODEL

While the discussion of the complete language specification is outside the scope of this book (as you can find it in the OMG specification document[29]), in the following we report a few excerpts of the specification in order to let the reader appreciate the flavor of the design and understand how the principles and requirements have been addressed in practice. Figure 7.15 shows an excerpt describing the main elements of the language. The main rationale here is to grant multiple abstraction levels: the generic `InteractionFlowElement` is an abstract metaclass never instantiated in the models, further specified by the concrete subclasses `ViewElement`, `ViewComponentPart`, `Action`, `PortDefinition`, and `Event`. The same applies to the `InteractionFlow` abstract metaclass, which is actually used in models only as either `NavigationFlow` or `DataFlow`. Notice that an `InteractionFlow` is a directed connection between two `InteractionFlowElements`, which in turn can have parameters. Figure 7.16 shows the detailed hiearchical structure of `ViewElement`, which is actually instantiated as either `ViewContainer` or `ViewComponent`: `ViewContainers` are then further extended in the Extensions package as example of UI containers, like Menu and Window (further examples may be screens, Web pages, panels, and so on). `ViewComponents` are extended as typical widgets that can be inserted inside `ViewContainers`,

---

[29]http://www.omg.org/spec/IFML

like List, Form, Details, and so on. Notice that a *composite* design pattern has been applied for describing the containment in an optimal way. A similar solution has been applied to the ViewComponentPart definition.

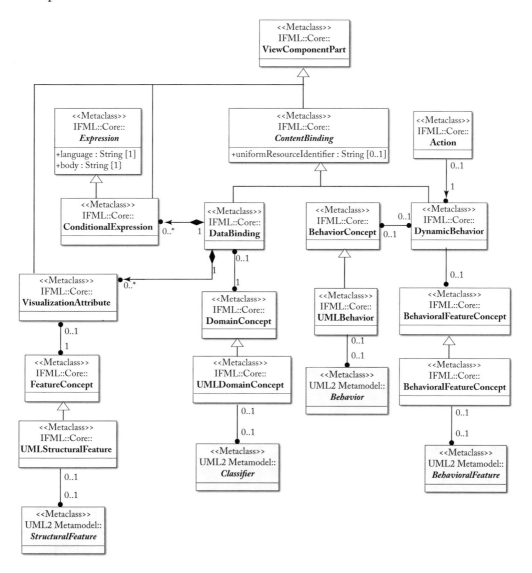

Figure 7.17: IFML Metamodel excerpt: ContentBinding and associated structure.

The metamodel excerpt in Figure 7.17 shows the definition of the integration with other models (and modeling languages). In particular, it describes how IFML connects to UML for the specification of the content model (defining the data to be included in the user interfaces) and for the specification of the business logic. Once again, multiple levels of abstraction are defined: the abstract concept of `ContentBinding` (used for connecting a `ViewComponent` to an external data source or a business logic) is extended by the actual possible bindings: `DataBinding` (that defines the reference `DomainConcept` for the component and the `VisualizationAttributes`) and `DynamicBehavior`. Again, through a multi-level abstraction mechanism, each alternative finally refers to a specific element of the UML metamodel to be referenced: `Classifier`, `StructuralFeature` (that is, a UML attribute or association end), `BehavioralFeature` (a UML method), or `Behavior` (a dynamic model in UML, such as a sequence diagram or an activity diagram).

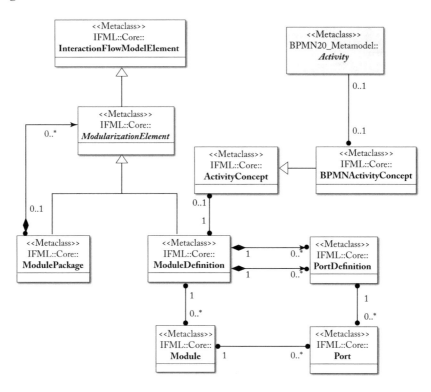

Figure 7.18: IFML Metamodel excerpt: Modularization.

Finally, Figure 7.18 describes how modularization and reuse is specified in IFML. Through a composite pattern, reusable modules can be defined as `ModuleDefinitions`, possibly aggregated in a hierarchy of packages. Module definitions have input and output ports for supporting parameter passing. Optionally, they can also be connected to a BPMN activity, meaning that the specified user interface module must be interpreted as the implementation of a specific business activity defined in a business process model. Once the module is defined, it can be instantiated multiple times inside models, through the `Module` metaclass and the associated `Port` metaclass.

7.5.5 IFML CONCRETE SYNTAX

In order to maximize acceptance and readability, a specific notation has been defined for IFML. The notation borrows as much as possible from existing widespread modeling languages, so as to keep coherence of meaning and symbols across notations.

For instance, event symbols are represented as circles, as customary in many languages (just think about UML activity diagrams, BPMN, or finite state machines). Containers are represented as squared boxes and components as rounded boxes. An example of the visual representation of an IFML model is given in Chapter 6, Figure 6.13b.

CHAPTER 8

Model-to-Model Transformations

Models are neither isolated nor static entities. As part of an MDE process, models are *merged* (to homogenize different versions of a system), *aligned* (to create a global representation of the system from different views to reason about consistency), *refactored* (to improve their internal structure without changing their observable behavior), *refined* (to detail high-level models), and *translated* (to other languages/representations, e.g., as part of code generation or verification/simulation processes).

All these operations on models are implemented as model transformations [61], either as Model-to-Model (M2M) or Model-to-Text (M2T) transformations (the latter is the topic of the next chapter). In the former, the input and output parameters of the transformation are models, while in the latter, the output is a text string. Analogously, Text-to-Model (T2M) transformations have a text string as input and a model as output; such transformations are typically applied in reverse engineering (cf. Chapter 3).

During the last decade, many efforts have been spent in designing specialized languages for specifying M2M transformations, ranging from textual to visual; declarative to imperative; and semi-formal to formal. We review most of them in the next sections and focus on two protagonists of different categories to illustrate the main characteristics of transformation languages.

8.1 MODEL TRANSFORMATIONS AND THEIR CLASSIFICATION

Since the introduction of the first high-level programming languages that were compiled to assembler, transformations are one of the key techniques in software engineering. Not surprisingly, model transformations are also crucial in MDE and come in different flavors to solve different tasks [19, 49].

In a general sense, an M2M transformation is a program which takes one or more models as input to produce one or more models as output. In most cases, *one-to-one* transformations, having one input model and one output model, are sufficient. As an example, consider the transformation of a class diagram into a relational model. However, there are also situations where *one-to-many*, *many-to-one*, or even *many-to-many* transformations are required, for instance, a model merge scenario where the goal is to unify several class diagrams into one integrated view.

Besides classifying model transformations based on the number of input and output models, another dimension is whether the transformation is between models from two different languages, referred to as *exogenous* transformations, or within models written with the same language which are then called *endogenous* transformations [49]. An example for an exogenous transformation is the typical MDA scenario where a platform independent model, e.g., a UML model, is transformed to a platform specific model, e.g., a Java model. A well-known example for endogenous transformations is model refactoring. Similar as for code, a model may be subject to quality improvements which may be achieved by restructuring the models using a transformation.

Furthermore, exogenous model transformations are not only usable for *vertical* transformations such as the aforementioned UML to Java scenario, where the abstraction level of the input and output models are different. Another usage is for *horizontal* transformations where the input and output models remain more or less on the same abstraction level. For instance, horizontal exogenous transformations are being used for realizing model exchange between different modeling tools, e.g., translating a UML class diagram to an ER diagram.

During the last decade, two central execution paradigms for model transformations emerged which are compared in Figure 8.1. First, there are *out-place* transformations for *generating* the output model from scratch (cf. Figure 8.1a). Such transformations are especially suited to exogenous transformations. Second, there are *in-place* transformations for *rewriting* a model by creating, deleting, and updating elements in the input model (cf. Figure 8.1b). Of course, this paradigm suits perfectly endogenous transformations such as refactorings.

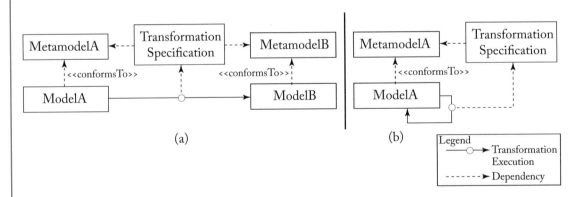

(a)   (b)

Figure 8.1: Different kinds of model transformations: (a) exogenous out-place vs. (b) endogenous in-place.

In the following sections, we present how to specify *exogenous* transformations as *out-place* transformations using ATL and *endogenous* transformations as *in-place* transformations using graph transformation languages.

8.2 EXOGENOUS, OUT-PLACE TRANSFORMATIONS

The ATLAS Transformation Language (ATL) [41] has been chosen as the illustrative transformation language for developing exogenous, out-place transformations, because it is one of the most widely used transformation languages, both in academia and industry, and there is mature tool support[1] available. ATL is a rule-based language which builds heavily on OCL, but provides dedicated language features for model transformations which are missing in OCL, like the creation of model elements (remember that OCL is side-effects free).

ATL is designed as a hybrid model transformation language containing a mixture of declarative and imperative constructs. ATL transformations are *uni-directional*, meaning that if a transformation from language A to language B is required, and vice versa, two transformations have to be developed. ATL transformations are operating on *read-only* source models and producing *write-only* target models. Please note that for out-place transformations, instead of *input model* the term *source model* is often used and instead of *output model* the term *target model* is often used. Thus, we use these terms interchangeably.

Coming back to the operation mode of ATL transformations, during the execution of a transformation, source models are queried but no changes to them are allowed. In contrast, target model elements are created, but should not be queried directly[2] during the transformation. The reason for both restrictions is that without them, the result of queries to source and target models may differ based on the execution state of the transformation which would be in contradiction to the nature of the declarative part of ATL.

Transformation Example.   For introducing the basics of ATL, we make use of our running example from the previous chapter. Assume we have to develop an excerpt of the transformation from sWML models to PSMs following the Model-View-Controller (MVC) pattern. For expressing the PSMs, a language for modeling MVC-based Web applications is applied, namely the simple MVC modeling language (sMVCML). Figure 8.2 illustrates two small excerpts of the source and target metamodels of this transformation.

Let us assume that the following two basic requirements have to be fulfilled by the transformation from sWML to sMVCML for the given metamodel excerpts. The requirements are graphically depicted in Figure 8.2 as class correspondences between the metamodel excerpts. While class correspondences act as a coarse-grained specification for a model transformation, a more fine-grained specification is given in the following in natural language also including value correspondences.

- **Requirement 1:** For each `Class` instance in the sWML model, a `Class` instance and a `DAO`[3] instance has to be created in the sMVCML model. The name of the sWML class should

---

[1]`http://www.eclipse.org/atl`
[2]We show later how to safely avoid this restriction.
[3]Data Access Object—`http://java.sun.com/blueprints/corej2eepatterns/Patterns/DataAccessObject.html`.

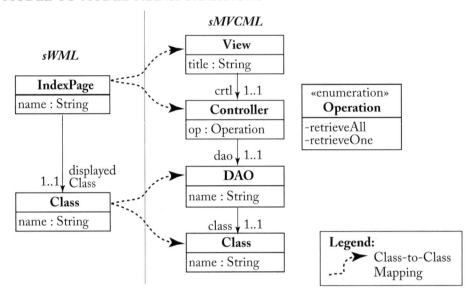

Figure 8.2: Metamodel excerpts of sWML and sMVCML.

become the name of the sMVCML class and the name of the DAO is the sWML class name concatenated with "DAO." Finally, the DAO has to comprise a link to the sMVCML class.

- **Requirement 2:** For each `IndexPage` instance in the sWML model, a `Controller` instance with assigned operation *showAll* as well as a `View` instance which contains the necessary UI elements (not shown in the following due to reasons of brevity) have to be generated in the sMVCML model. The `View` instance has to be linked to the `Controller` instance. Furthermore, the `Controller` instance has to be linked to the `DAO` instance produced for the sWML `Class` instance which is linked by the `IndexPage` instance via the `displayedClass` reference.

Listing 8.1 shows the transformation implementing these two requirements in ATL. This listing is now used to explain the ATL language features by example.

Anatomy of ATL transformations.   A transformation defined in ATL is represented as a *module* divided into a *header* and a *body* section. The header section of an ATL transformation states the name of the transformation module and declares the source and target models which are typed by their metamodels. There can be more than one input model and output model for an ATL transformation.

In Listing 8.1, one input model of type sWML (cf. **from** keyword) and one output model of type sMVCML (cf. **create** keyword) is defined. Thus, the scenario is a typical *one-to-one* model transformation.

The body of an ATL transformation is composed by a set of *rules* and *helpers* which are stated in arbitrary order after the header section.

- **Rules:** Each rule describes how (a part of) the target model should be generated from (a part of) the source model. There are two kinds of declarative rules[4] in ATL: *matched* rules and *lazy* rules. The former is automatically matched on the input model by the ATL execution engine, whereas the latter has to be explicitly called from another rule, giving the transformation developer more control over the transformation execution.

  Rules are mainly composed of an *input* pattern and an *output* pattern. The input pattern filters the subset of source model elements that are concerned by the rule by defining one or more input pattern element(s). In particular, an obligatory type (corresponding to the name of a metaclass defined in the source metamodel) has to be stated for each input pattern element as well as an optional filter condition, expressed as an OCL expression. The type and the filter constrain on which model elements the rule should be applicable. The output pattern details how the target model elements are created from the input. Each output pattern element can have several *bindings* that are used to initialize the features of the target model elements. The values assigned in the bindings are calculated by OCL expressions.

- **Helpers:** A helper is an auxiliary function that enables the possibility of factorizing ATL code used in different points of the transformation. There are two kinds of helpers. First, a helper may simulate a derived attribute that needs to be accessible throughout the complete transformation. Second, a helper may represent an operation which calculates a value for a given context object and input parameters. Opposed to rules, helpers cannot produce target model elements, they can only return values which are further processed within rules.

In Listing 8.1, we have defined two matched rules to implement the aforestated requirements. The first rule is responsible for transforming instances of sWML!Class into instances of sMVCML!Class and sMVCML!DAO. Thus, this rule has a simple input pattern (cf. keyword **from**) matching for all instances of the metaclass sWML!Class. The output pattern of this rule (cf. keyword **to**), creates for each match, an sMVCML!Class instance and an sMVCML!DAO instance by using two corresponding output pattern elements. The bindings of the output pattern elements are straightforward. The names of the input pattern elements (e.g., c1) are used to access the input pattern elements in the bindings of the output pattern elements to retrieve necessary information of the input model to build the output model. By this, the names of the classes contained in the input model can be transferred to the output model (name <- c1.name). Furthermore, output pattern elements may be directly linked in the output pattern by using again the names of the output pattern elements in the bindings, e.g., class <- c2.

---

[4]To be precise, there is also a third kind of ATL rules, namely *called* rules which are very similar to lazy rules.

For transforming `IndexPages` instances into `Controller` and `View` instances, the second rule is introduced. This rule has again a simple input pattern having only one element and an output pattern comprising again two elements. The bindings of the output pattern elements are again straightforward, except the binding for the reference `Controller.dao`. The reason why this binding is more complex is that a `Controller` instance has to refer to a `DAO` instance which is created by another rule. Thus, we do not have a direct pointer to this target element as before when linking the `DAO` and the `Class` instances which are both created in the first rule. To link to a target element which is created by a different rule, we have to resolve the target element for a given source element. For this, ATL provides a specific operation called `resolveTemp` which is able to retrieve produced target elements for a given source element. Thus, for linking the `Controller` to the `DAO`, we make use of this operation to get the produced `DAO` instance for the `Class` instance which is linked by the `IndexPage`. Please note that for simple cases the `resolveTemp` operation does not have to be used explicitly. In such cases the element produced by the first output pattern element is automatically retrieved. More information on the `resolveTemp` operation is presented in the following part which describes the execution phases of ATL. These phases have a major impact on the point in time when target elements are actually resolved. In contrast, linking instances which are produced by the same rule is much simpler. Only the variable of the output pattern element has to be used as value for the binding (e.g., `crtl <- c`). Please be aware that for computing the title of `View` instances, a dedicated helper is introduced which is called in the binding for setting the title attribute. This helper may be seen as a derived attribute for the `IndexPage` class.

Listing 8.1: ATL Transformation Example

```
module SWML2MVC;
create OUT : sMVCML from IN : sWML;

-- implementation of requirement 1
rule Class2Class_DAO {
  from
    c1 : sWML!Class
  to
    c2 : sMVCML!Class (
      name <- c1.name
    ),
    d : sMVCML!DAO(
      name <- c1.name + 'DAO',
      class <- c2
    )
}

-- implementation of requirement 2
rule IndexPage2Controller_View {
  from
    p : sWML!IndexPage
  to
    c : sMVCML!Controller (
      op <- #retrieveAll,
      dao <- thisModule.resolveTemp(p.displayedClass, 'd')
    ),
    v : sMVCML!View(
      title <- p.title,
      ctrl <- c
    )
```

```
}
helper context sWML!IndexPage def : title : String =
  'Show all ' + self.displayedClass.name + ' entries';
```

Execution phases of ATL transformations. ATL is a powerful language which allows us to define transformation in a concise syntax. To explain the details hidden behind the ATL syntax, we now take a look on the actual transformation execution achieved by the ATL virtual machine. The execution of an ATL transformation is structured into three sequential phases which are explained in the following and illustrated using a small excerpt of a concrete transformation run (cf. Figure 8.3) of the ATL transformation shown in Listing 8.1.

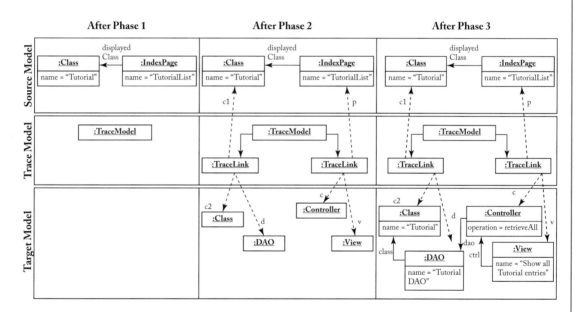

Figure 8.3: ATL transformation execution phases by example.

Phase 1: Module initialization. In the first phase, among other things, the *trace model* for storing the *trace links* between source and target elements is initialized. In the subsequent phase 2, each execution of a matched rule will be stored in the trace model by creating a trace link pointing to the matched input elements and to the created output elements. As we will see later, the trace model is an important concept for exogenous transformations: (i) to stop the execution of a transformation and (ii) to assign features of the target elements based on values of the source elements.

Example: In our example, simply the trace model is initialized in this phase as illustrated in the left-hand side of Figure 8.3.

Phase 2: Target elements allocation. In the second phase, the ATL transformation engine searches for matches for the source pattern of the matched rules by finding valid configurations of source model elements. When the matching condition of a matched rule (all input pattern elements are bound and the filter condition is valid) is fulfilled by a configuration of source model elements, the ATL transformation engine allocates the corresponding set of target model elements based on the declared target pattern elements. Please note that only the elements are created, but their features are not yet set. Furthermore, for each match, there is a trace link created which interconnects the matched source elements and the generated target elements.

Example: In our example, as is shown in the middle of Figure 8.3, there is one match for both rules. The target elements are allocated in the target model and the trace links are created and attached to the trace model.

Phase 3: Target elements initialization. In the third phase, each allocated target model element is initialized by executing the bindings that are defined for the target pattern element. In the bindings, invocations of the *resolveTemp* operation are quite common. This operation allows for the reference to any of the target model elements which have been generated in the second execution phase for a given source model element. It has the following signature: `resolveTemp(srcObj:OclAny,targetPatternElementVar:String)`. The first parameter represents the source model element for which the target model elements have to be resolved. The second parameter is the variable name of the target pattern element which should be retrieved. The second parameter is required, because it is possible to generate several target elements for one source element by using multiple target pattern elements in one rule as it is the case in our example. Thus, when a source model element has to be resolved, the variable of the target pattern element which has produced the requested target element has to be given.

Example: In our example, some bindings are simple, meaning that only target model elements produced within one rule are linked or the matched source model elements contain directly the information (such as attribute values) needed for setting the features of the target model elements. However, there is one binding which makes use of the *resolveTemp* operation (`resolveTemp(p.displayedClass,'d')`). Before the *resolveTemp* operation is executed, the displayed class for the matched page is queried by evaluating the first parameter of the operation. Subsequently, by executing the *resolveTemp* operation, the trace link for the displayed class is automatically retrieved, and finally, the variable names of the output pattern elements are used to retrieve the actual requested target element (specified by the second parameter of the operation) from the set of linked target elements (`{'c2','d'}`) of the trace link. You can think of retrieving the output element by navigating from the trace link to the output element by using the output pattern element variable as reference name.

Internal vs. external trace models.   The above presented trace model is internal to the ATL transformation engine. It is the key for setting features of target elements in a convenient way and for stopping the rule execution since a rule is only executed for a match by the ATL execution engine, if this match, i.e., the set of matched input elements, is not already covered by an existing trace link.

If for external traceability reasons, a persistent trace model is needed, e.g., to see the possible impact of source model changes on the target model, this internal trace model (which is by default only transient) may be persisted as a separate output model of the model transformation [40, 72].

Alternatives to ATL.   Besides ATL, there are other dedicated transformation approaches for exogenous, out-place transformations.

QVT. The Query-View-Transformation (QVT) standard of the OMG covers three languages for developing model transformations. First, the QVT Relational language is a declarative approach to define correspondences between metamodels. However, in contrast to ATL, these correspondences are not defining a transformation direction, instead they can be interpreted bi-directionally. This allows for the derivation of transformations for both directions as well as to check the consistency between two models and even synchronize two models in case they are not consistent (a more detailed description of these terms is provided in SubSection 8.4.4). Second, the QVT Operational language is an imperative approach for developing model transformations in a uni-directional way. There is also out-of-the-box support for tracing, but the dispatching of transformations rules has to be defined by the user, just as lazy rules are used in ATL. Third, the QVT Core language is a low-level language which is designed to be the target of the compiler of the QVT Relation language. Thus, it is not intended to write transformations directly in QVT Core. There is tool support in Eclipse for the declarative languages[5] (QVT Relations and QVT Core) as well as for QVT Operational.[6]

TGG. Triple Graph Grammars (TGG) [60] are a dedicated approach for defining a correspondence graph between two metamodels, which allows us to transform models in both directions. Furthermore, the correspondence graph may be used to synchronize models and to check if they are consistent. Thus, TGGs are targeting similar scenarios as QVT Relations, but provide a strong theoretical basis. There is tool support for TGGs in MOFLON[7] and in Eclipse by the TGG Interpreter[8] and by the TGG tool available at MDElab.[9]

ETL. In the Epsilon project, several languages have been developed for model management. The Epsilon Transformation Language[10] (ETL) supports exogenous, out-place model transformations similar to ATL, but it provides also additional features, such as the possibility to modify the input model elements during transformations.

---

[5]http://wiki.eclipse.org/M2M/QVT_Declarative_(QVTd)
[6]http://wiki.eclipse.org/M2M/Operational_QVT_Language_(QVTO)
[7]http://www.moflon.org
[8]http://www.cs.uni-paderborn.de/index.php?id=12842&L=1
[9]http://www.mdelab.de
[10]http://www.eclipse.org/epsilon/doc/etl

8.3   ENDOGENOUS, IN-PLACE TRANSFORMATIONS

In what has been discussed until now, models were manually modified by applying actions in modeling editors such as adding and removing model elements or updating their values. However, there is a plethora of scenarios where modifications of models should or have to be automated. Recall that out-place transformations require us to completely build the output model from scratch. If out-place transformations were to be used for introducing modifications to a model, this would require copying the complete source model to the target model, except those elements to be deleted or modified. Therefore, alternative execution modes for model transformations are available that are a better fit for this transformation scenario. For instance, *in-place* transformations provide changes to a model without copying the static parts of the model, i.e., the elements which are not touched by the transformation. Thus, only transformation rules taking care of the dynamic part, i.e., the changes (but no rules for simply copying unmodified elements) are required.

Graph transformation [21] is a very elegant approach to implement in-place model transformations. Thus, they have been selected for demonstrating the development and usage of in-place model transformations.

Graph Transformation Basics.   Graph transformation is a declarative, rule-based technique for expressing in-place model transformations based on the fact that models and meta-models can be expressed as graphs (with typed, attributed nodes and edges). Thus, models can be manipulated by graph transformation techniques.

Graph transformations are especially useful to define in-place transformations to support, e.g., model simulation, optimization, execution, evolution, and refactoring. Nevertheless, they are so general that also out-place transformations could be formulated with graph transformations by representing the source and target models as well as the trace model as one integrated graph.

Graph transformations are popular due to their visual form (making rules intuitive) and formal nature (making rules amenable to analysis). For example, graph transformations can be used to describe the operational semantics of modeling languages for implementing a model execution engine, providing the advantage that it is possible to use the abstract syntax and sometimes even the concrete syntax of the modeling language to define the transformation rules, which then become very intuitive to the designer.

A *graph grammar* consists of a set of graph transformation rules and an initial graph (often referred to as *host graph*) to which the rules are applied. The core of a rule comprises a left-hand side (LHS) graph and right-hand side (RHS) graph. The LHS expresses pre-conditions for the rule to be applied, whereas the RHS contains the rule's post-conditions. The actions that are going to be carried out by the rule are implicitly defined by both sides. More precisely, the execution of a transformation rule produces the following effects: (i) all elements that only reside in the LHS are deleted; (ii) all elements that only exist in the RHS are added; and (iii) all elements that reside in both sides are preserved. To mark that an element in the RHS is equivalent to an element in the LHS, the elements must have the same identifier assigned.

In order to apply a rule to the initial graph, a *morphism* (often also referred to as *occurrence* or *match*) of the LHS has to be found in it. If several matches are found, one is selected randomly. Subsequently, the rule is applied for the selected match by substituting the match by the RHS. The grammar execution proceeds by applying the rules in non-deterministic order, until none of them is applicable. There are also other forms of graph transformation systems, which allow for more control over the actual transformation execution. More details about these forms are presented later in this section.

Graph transformation rules may be applied fully automated by the graph transformation engine as long as possible in arbitrary order. However, there are also cases which require a more interactive execution mode. Assume there should be a refactoring applied on a specific model element. Thus, this element has to be selected by the user and represents an important input parameter for the transformation rule which implements this refactoring. Thus, some graph transformation approaches allow for pre-bound LHS elements by providing explicit input parameters which can be set by the user before executing transformation rules. Furthermore, additional user input may be required during the execution of a transformation rule, e.g., for setting the values for features of model elements which cannot be derived from the current model state. Consequently, some approaches allow us to query user input during transformation execution.

Transformation example.   To exemplify the usage of graph transformations, we make use of the following transformation example. For speeding up the development of sWML models, there should be a kind of model completion support which allows us to instantly introduce several elements in the hypertext layer for a selected class in the content layer. In particular, an `IndexPage` (reachable from the homepage of the Web application) for showing a list of all instances of the class should be created. Additionally, this page should be linked to a `DetailsPage` displaying the details of one selected instance. Such model completion operations are a typical example for using in-place transformations. The transformation rule for introducing the mentioned pages for a given class is shown in Figure 8.4.

The LHS of the rule states that there has to exist a class (which has to be selected by the user by setting the input parameter of the transformation rule) as well as a hypertext layer with a home page already contained. The hypertext layer is needed to include the pages to be generated. These pages are stated only in the RHS of the rule, which means that they are created by the rule's execution. Please note that this rule is not deleting elements, thus the LHS is fully contained in the RHS. The attribute values for the new pages are computed by using expressions which also access properties of other elements by using their variable names (e.g., `name = c.name + "List"`). Normally, these expressions are defined in OCL or using scripting languages.

A major concern is to control the application of rules. The LHS of a rule specifies what must exist in a graph to execute the rule. However, often it is required to describe what must not exist in a graph to apply a rule. Therefore, Negative Application Conditions (NACs) have been

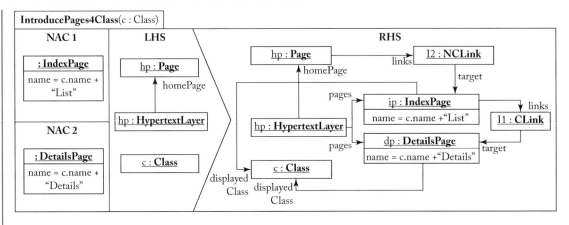

Figure 8.4: Example graph transformation rule: *Introduce Content Pages for Selected Class.*

introduced for graph transformations. A NAC is a graph which describes a forbidden sub graph structure in the initial graph, i.e., the absence of specific nodes and edges must be granted. A graph transformation rule containing an NAC is executed when a match for the LHS is found and the NAC is not fulfilled for this match. Not only can one NAC be attached to one rule, but several. In our example, two NACs are used to avoid re-introducing already existing IndexPages and DetailsPages. Please note that for this example, two NACs are necessary and not only one. Having only one NAC would only forbid the application of the rule if an IndexPage and a DetailsPage exist, but not if only one of them exists.

To exemplify the execution of the presented graph transformation rule, consider Figure 8.5. The Class *Tutorial* is selected by the user and the transformation rule is executed on graph G. The pre-condition of the rule is fulfilled, because there exists a hypertext model with a homepage and we assume that no IndexPage and DetailsPage exist for the given class. Thus, the RHS rewrites the graph G, in particular, in the hypertext model, two pages (*TutorialIndex*, *TutorialDetails*) which are appropriately linked are created. Please note that in the resulting graph G', the RHS of the transformation rule can be completely matched.

Advanced graph transformations techniques.   There are several advanced techniques for specifying, executing, and analyzing graph transformations.

Alternative notations. Until now, transformation rules have been defined in the abstract syntax of the modeling language. An even higher readability may be reached by using the concrete syntax of the modeling languages for defining the transformation rules as shown for our graph transformation rule example in Figure 8.6. However, only a few graph transformation tools provide this capability. Furthermore, there are some approaches that use a condensed graphical notation by merging the NACs, LHS, and RHS into one graph and using some annotations

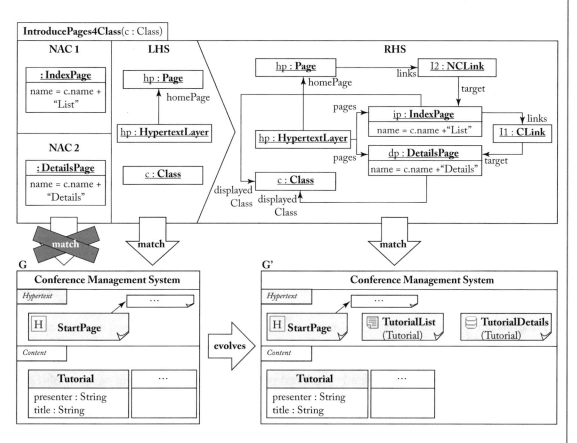

Figure 8.5: Application of the graph transformation rule of Figure 8.4.

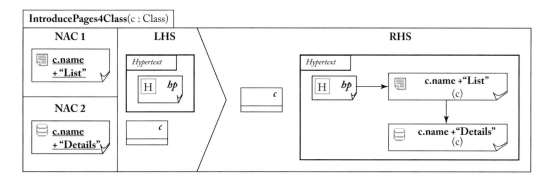

Figure 8.6: Graph transformation rule of Figure 8.4 shown in concrete syntax.

for marking forbidden, preserved, deleted, or created elements. Finally, some approaches use a textual concrete syntax to define the graph transformation rules. We discuss protagonists of these different categories in the end of this section.

Rule scheduling. A graph transformation system consists of a set of different graph transformation rules. Now the question arises in which order they are executed. If nothing else is specified, non-deterministic selection of rules is assumed until no rule matches anymore, as we have discussed for graph grammars. If there are different rule execution sequences, several different output models may be possible in theory. However, there are also cases where a deterministic execution of a graph transformation system is preferred. For supporting such cases, two extensions have been introduced. First, a basic approach is to define *priorities* for rules, meaning that if several rules are possible to be matched at any point in time, the rule with the highest/lowest (depends how priorities are actually ordered in the different approaches) priority is executed. Second, a more advanced approach is to allow for a powerful *orchestration* of graph transformation rules by using programming-like control structures, often referred to as graph transformation units, e.g., for defining loops, conditional branches, and so on. These approaches are summarized under the term *programmable graph transformations* [66].

Analysis. Because graph transformations are a declarative approach and based on a strong theoretical basis, there are several analysis methods for graph transformation systems available. First, in case non-deterministic graph transformation systems are employed, there is the question if always the same unique model is finally produced when the rules are applied in any possible order. For this case, critical pairs between the rules can be computed [32]. A critical pair is defined as two rules being mutually exclusive, i.e., an application of rule 1 hinders or enables the application of rule 2. Finally, some graph transformation approaches allow us to reason about the termination of a graph transformation system.

Tool Support.   There are several Eclipse projects available which specifically focus on providing graph transformation technologies for EMF. However, besides this common goal, the different projects support diverse features for implementing and executing graph transformations. Thus, the question of which graph transformation tool to use has to be decided case-by-case depending on the actual transformation problem.

Henshin[11] is the successor of EMF Tiger and aims at introducing several advanced features such as programmable graph transformations and several analysis features such as model checking support. Furthermore, an integration with AGG[12] is available which allows us to use several advanced graph transformation techniques such as computing critical pairs between transformation rules.

---

[11]http://www.eclipse.org/modeling/emft/henshin
[12]http://user.cs.tu-berlin.de/~gragra/agg

Fujaba[13] is a protagonist of programmable graph transformation approaches. In particular, story diagrams (comparable to UML activity diagrams) are used to orchestrate graph transformation rules. Within the graph transformation rules, Java is used for defining application conditions for the LHS and assignments for the RHS.

e-Motions[14] is an Eclipse plugin to graphically specify the behavior of modeling languages by using graph transformation rules shown in the graphical notation of the modeling languages. One unique feature of e-Motions is the possibility to specify time-related attributes for rules, e.g., duration or periodicity. Having defined the behavior of a modeling language in e-Motions, the models can be simulated and analyzed by translating the models and the model transformation to Maude[15] which is a programming framework based on rewriting logic.

ATL Refining. *ATL Refining* [68] adds a new execution mode to ATL tailored for endogenous in-place transformations. The ATL Refining mode is activated by simply substituting the *from* keyword with the *refining* keyword in the header of an ATL transformation. In-place transformation rules are syntactically written as standard out-place rules comprising an input pattern and an output pattern. However, the execution of the rules is different. If an output pattern element is already defined as an input pattern element (by sharing the same variable name), the element is updated by the bindings, but not created. Newly introduced output pattern elements (which have no counterpart in the input pattern) are generated by the rule. For deleting elements, an output pattern element can be annotated with a special keyword *drop* which means the element will be deleted with all its containees, i.e., all elements directly or indirectly contained by this element. To summarize, ATL Refining represents an in-place transformation language using a textual syntax.

8.4   MASTERING MODEL TRANSFORMATIONS

In this section, we give a short overview and references to additional information on advanced transformation techniques which are especially suited to deal with the complexity of model transformations in practical settings.

8.4.1   DIVIDE AND CONQUER: MODEL TRANSFORMATION CHAINS

Transformations may be complex processes which should be modeled themselves and structured into different steps to avoid having one monolithic transformation. Transformation chains are the technique of choice for modeling the orchestration of different model transformations. Transformation chains are defined with orchestration languages that enable us to model the sequential steps of transformations. This means the input models for the first transformation are defined, the first transformation is selected, the output models of the transformation become the input

---

[13]http://www.fujaba.de
[14]http://atenea.lcc.uma.es/index.php/Main_Page/Resources/E-motions
[15]http://maude.cs.uiuc.edu

models for the second transformation, and so on. More complex transformation chains may also incorporate conditional branches, loops, and further control constructs.

Transformation chains are not only a means for splitting the transformation to be developed into several modules, but also allow the construction of complex model transformations from already defined transformations. Furthermore, having smaller transformations focusing on certain aspects may also allow for higher reusability.

Model transformation chains may be defined in well-known textual build languages such as ANT as it is possible for ATL.[16] However, there are also dedicated languages to graphically model transformation chains by using a subset of the UML activity diagram language, e.g., the Wires*[17] approach allows us to graphically orchestrate ATL transformations.

The orchestration of transformations directly leads to the research field of modeling in the large. We come back to this topic in Chapter 10 where model management tasks are discussed, whereas these tasks may be defined as transformation chains.

8.4.2   HOT: EVERYTHING IS A MODEL, EVEN TRANSFORMATIONS!

So far, we have seen transformations as operations to manipulate models, but in fact, transformations can be considered as models themselves, since they are instances of a transformation metamodel, i.e., an ATL transformation can be expressed as an instance of the ATL metamodel (defining the abstract syntax of the ATL language), and thus, can be seen as a model.

This uniformity allows us to reuse tools and methods, i.e., the same tools that can be used to create models can be used to create transformation models, and it creates a framework that can in theory be applied recursively since transformations of transformations can be transformed themselves. Moreover, it is important to facilitate the manipulation of transformations. Just as a normal model can be created, modified, and augmented through a transformation, a transformation model can itself be created, modified, and so on, by a so-called *Higher Order Transformation* (HOT) [67]. This means that we can write transformations that take as input a model transformation and/or generate a model transformation as output. For instance, we could write a HOT that automatically refactors a set of transformations to improve their internal structure. Another example would be to add a logging aspect to the transformation.

8.4.3   BEYOND BATCH: INCREMENTAL AND LAZY TRANSFORMATIONS

Until now we have elaborated on one execution strategy for out-place model transformations: read the complete input model and produce the output model from scratch by applying all matching transformation rules. Such executions are often referred to as *batch* transformations. However, two scenarios may benefit from alternative execution strategies: (i) an output model already exists from a previous transformation run for a given input model and (ii) only a part of the output model is needed by a consumer, such as a subsequent transformation or modeling editor. For

---

[16]http://wiki.eclipse.org/ATL/User_Guide_-_The_ATL_Tools/#ATL_ant_tasks
[17]http://atenea.lcc.uma.es/index.php/Main_Page/Resources/Wires*

the former scenario, *incremental* transformations [42] may be employed, while for the latter *lazy* transformations [68] are of special interest.

Incremental transformations, as their name suggests, do not transform the complete input model to a new output model created from scratch each time they are executed. Instead, they only consider the differences between the current input model and the input model used in the last transformation run to minimize the changes to be applied on the output model, e.g., if a new model element is added in the input model, only the rules that are a match for the new element will be executed to update the output model. The advantages of this transformation strategy are two-fold: (i) the transformation of the changes is normally more efficient than translating the complete state of the model, and (ii) changes applied on the output model are preserved.

Lazy transformations are driven by access operations on the target model. As known from programming languages which employ lazy evaluation for computing expressions only as far as required in a given point in time, target models also may be produced only under request. For example, assume the only consumer of the output model is another transformation. Imagine that the second transformation never accesses some elements of the target model produced by the first one. Clearly, then, creating those elements in the first place would be a waste of time since nobody will ever read them. Another example is when a certain feature of a model element is only very lately accessed in the execution of the second transformation. Thus, the evaluation of the binding used in the first transformation to calculate this feature may be deferred to this point in time instead of computing upfront. The advantages of this transformation strategy are two-fold: (i) as for incremental transformations, the execution of the transformation is more efficient for certain cases, and (ii) the transformation of very large models (even becoming infinite models) is enabled.

8.4.4   BI-DIRECTIONAL MODEL TRANSFORMATIONS

Until now, we have seen uni-directional model transformations languages—except the short discussion on QVT and TGG when mentioning some alternatives to ATL. In the last decade, effort has been made to develop bi-directional model transformation languages [33, 65] which (i) do not impose a transformation direction when specifying a transformation and (ii) allow for different execution modes such as *transformation*, *integration*, and *synchronization*.

The transformation mode is further divided into forward transformation and backward transformation. Forward transformations allow us to go from source models to target models, while backward transformations do the inverse. Such transformations may also be specified by two uni-directional transformations, but the strong advantage of using a bi-directional transformation is that the consistency between forward and backward transformation is given by construction [33].

The integration mode assumes to already have the source and target models given and computes the correspondences between source and target model elements. This means, instead of generating elements in the one model for a given match of a transformation rule in the other model, the integration mode checks if the expected elements (that would be produced in the transformation mode) exist in the other model. If for every match in the source model the corresponding elements in the target can be found (and vice versa), the two models are considered to be fully integrated. This shows quite well that bi-directional transformations may be seen as constraints that have to be fulfilled by a pair of models. If all constraints are fulfilled, i.e., all correspondences can be established, the two models are fully integrated.

The synchronization mode starts with the source and target models and the computed correspondences by using the integration mode. More precisely, if two models are not fully integrated, the synchronization mode aims at establishing a full integration by applying some of the defined transformation rules. For example, if in one model a new element has been added, we may find an additional match of a rule for this element. The synchronization mode now executes this rule for the additional match to produce the corresponding elements on the other side. Another example would be that an element is deleted in one of the models and now a match is no longer valid. Consequently, the corresponding elements on the other side have to be deleted as well.

CHAPTER 9

Model-to-Text Transformations

Several concepts, languages, and tools have been proposed in the last decade to automate the derivation of text from models by using Model-to-Text (M2T) transformations. Such transformations have been used for automating several software engineering tasks such as the generation of documentation, task lists, etc.

Of course, the killer application of M2T transformations is code generation. Actually, the main goal of model-driven software engineering is at the end of the day to get a running system out of its models. Current execution platforms are mostly code based—with a few exceptions which allow for a direct interpretation of models (cf. discussion in Chapter 3). Thus, M2T transformations are mostly concerned with code generation to achieve the transition from the model level to the code level. Please note that not only the code representing the system to develop may be derived from the models, but also other code-related artifacts such as test cases, deployment scripts, etc. In addition, formal code descriptions may be derived which allows us to analyze different properties of a system (cf. Section 10.7). This is one of the major benefits of using models—they may be used constructively to derive the system as well as analytically to better explore or verify the properties of systems. M2T transformations are currently the bridge to execution platforms and analysis tools.

In this chapter, we start with the basics of model-based code generation using M2T transformations, discuss different approaches to implement M2T transformations, and finally, report on techniques for mastering the complexity of code generators.

9.1 BASICS OF MODEL-DRIVEN CODE GENERATION

Code generation has a long tradition in software engineering going back to the early days of high-level programming languages where the first compilers developed. However, there is a difference in the goal of code generation in compilers [2] and in MDE. While in the context of compilers, code generation is the process of transforming source code into machine code, in MDE, code generation is the process of transforming models into source code. Thus, MDE code generation is built on top of existing compilers for programming languages.

The following three questions are essential when one has to develop a model-based code generator.

- **How much is generated?** The main question is which parts of the code can be automatically generated from models. Is a *full* or only a *partial* code generation possible? Partial can mean several things in this context. First, it can mean that one (horizontal or vertical) layer of the application is completely generated while another layer may be completely manually developed. Furthermore, it can also mean that one layer is only partially generated and some missing parts have to be manually completed. Partial code generation may also refer to the model level, by using only code generation for certain parts of a model while other parts are not touched by the code generator and have to be implemented manually.

- **What is generated?** It is important to know what kind of source code to generate. Of course, the code to be generated should be as concise as possible, and please keep in mind the Turing test for code generators mentioned in Chapter 3 emphasizing the need for generating code developers are able to read. Therefore, current high-level programming languages, APIs, and frameworks should be exploited as much as possible for avoiding to reinvent the wheel when developing a code generator. Choosing the right programming paradigm is the first decision to make. The motto here is: the less code to generate which is able to represent a system, the better.

- **How to generate?** Once the requirements for the code generation are specified, namely *what* has to be achieved by the code generator, i.e., which parts are generated and which target languages are used, it has to be decided *how* to implement these requirements. As we shall see later in this chapter, several languages may be employed to generate code from models, ranging from GPLs to DSLs.

Code generation may be described as the vertical transition from models on a higher-level of abstraction to lower-level artifacts. The benefits of MDE are heavily based on this vertical transition. Thus, code generators have to bridge this gap in the abstraction which may be achieved in different ways.

Closing the gap between models and code.   When a modeling language is employed to build the models, normally not all specifics of the underlying technologies may be expressible, since a key goal in MDE is to abstract from technology details. For example, assume you are defining a UML class diagram for a specific domain and you have to introduce several attributes of type String. If you want to deploy this class diagram in an environment with limited space for storing data, you may want to define a restricted length for these attributes instead of using the maximum possible String length. Thus, such as in the MDA approach, additional information may be needed later on in order to reach the goal of having executable software.

Information has to be either provided by the modeler by using model augmentations such as UML profiles for specific technologies, by applying the convention-over-configuration principle for the code generation, or by leaving the specification open on the model level and fill the details at the code level. The first approach allows one to tweak several details for the derived implementation by spending more effort in preparing models for code generation. The second approach does not need this additional effort, but optimizations of the derived implementation may only be done directly on the code level. Of course, a hybrid approach by starting with convention-over-configuration in combination with manual tweaking by model augmentations may be also a way to go. Using the third approach, only partial implementations are generated which have to be completed by the developer on the code level. Here, special care has to be taken, because having models and code on which changes are applied contradicts the single source of information principle. Countermeasures may be introduced such as so-called protected areas within the code which are special areas requiring for manual written code or other explicit extension points which have to be filled by the developers, e.g., having base classes generated which have to be extended by user-defined classes for implementing missing functionality which cannot be specified on the model level.

9.2   CODE GENERATION THROUGH PROGRAMMING LANGUAGES

In general, the implementation of a code generator can be based in its turn on MDE principles or just employ a more traditional programming approach. Following the latter approach, a code generator may be implemented as a program using the model API automatically generated from the metamodel to process the input models and print out code statements to a file using standard stream writers provided by APIs of the programming language used to implement the code generator.

The model API is realized in EMF by using itself an M2T transformation which reads an Ecore-based metamodel and generates a Java class for each Ecore class. In Figure 9.1, there is an excerpt from the sMVCML metamodel shown on the left-hand side and the corresponding Java classes on the right-hand side. The mapping from Ecore to Java is mostly straightforward—this was one of the design goals of Ecore. For each feature of the metaclasses, corresponding getter and setter methods are generated on the Java side. This means, a model can be read, modified, and completely created from scratch by using this generated Java code instead of using modeling editors. For more information on how to use the generated model APIs and the powerful EMF API we refer the interested reader to [64].

Before we delve into specific M2T transformation languages in Section 9.3, we present how a GPL may be employed to develop a code generator. By doing so, we demonstrate: (i) how models are processed using a model API generated from the metamodel and (ii) the features needed to realize a code generator.

As illustrated in Figure 9.2, the following phases have to be supported by a code generator.

1. **Load models:** Models have to be deserialized from their XMI representation to an object graph loaded in-memory. For this, current metamodeling framework APIs provide specific operations.

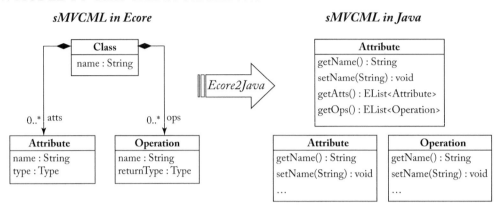

Figure 9.1: Generated model API from an excerpt of the sMVCML metamodel.

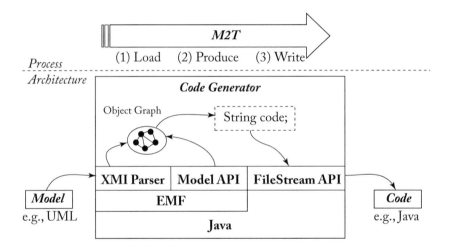

Figure 9.2: Code generation through programming languages: Java code that generates Java code.

2. **Produce code:** Collect the model information needed for generating the code by using the model API to process the models. Typically, the object graph is traversed starting from the root element of a model down to its leaf elements.

3. **Write code:** Code is, for instance, saved in String variables, and finally, persisted to files via streams.

Code generation example.   In the previous chapter, we have shown how sWML models are translated into sMVCML models. The sMVCML models have to be subsequently translated to code to end up with an executable Web application. An excerpt of this M2T transformation is now used to show how code generators are implemented. Figure 9.3 shows an exemplary translation of a sMVCML class (please note that for sMVCML, the UML class diagram notation is reused) of our running example (cf. left-hand side) into corresponding Java code (cf. right-hand side). As can be seen in this figure, the translation is straightforward, but sufficient to show the major concepts required to implement code generators. The sMVCML class is translated into a Java class which implements the `Serializable` interface, the sMCVML attribute into a private variable with getter/setter methods for accessing/modifying the variable, and the sMVCML operation into a Java method. However, concerning the latter, only the method signature can be derived from the sMVCML model. The implementation of the methods is postponed to the code level. Thus, a partial code generation approach is used. A major requirement is to take care of manually added code within automatically generated code to allow for an iterative model-driven development process.

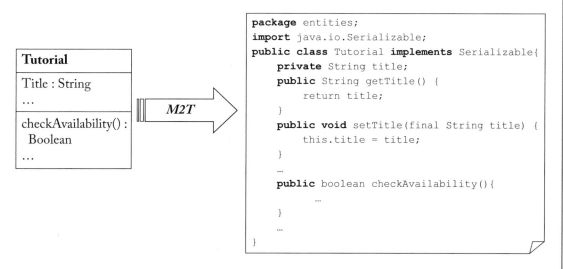

Figure 9.3: Excerpt of sMVCML model and corresponding code.

The Java program for producing the discussed Java code from the sMVCML language is shown in Listing 9.1. For the first phase, namely loading the models, the EMF API, providing classes for loading resources, in our case the sMVCML models, into memory, is used. In phase two, all model elements are queried from the input model, and subsequently, iterated. If the model element is a class (cf. `instanceof` type check), a String variable named `code` is initialized and further filled with Java statements as String values. In phase three, a stream is defined to a Java file with the name of the processed class and the value of the `code` variable is persisted to this file. Of course, more sophisticated GPL-based code generators may be developed using design patterns such as the Visitor pattern [27], but the drawbacks stated in the following also apply for such solutions.

Listing 9.1: Java-based Code Generation

```
// PHASE 1: load the sMVCML model using the EMF API
ResourceSet resourceSet = new ResourceSetImpl();
Resource resource =
resourceSet.getResource(URI.create("model.smvcml"));

// PHASE 2: collect the code statements in variable
// traverse the complete model using the EMF API
TreeIterator allElementsIter = resource.getAllContents();
while (allElementsIter.hasNext()) {
    Object object = allElementsIter.next();
    if (!object instanceof Class) continue;
    Class cl = (Class) object;

    // String variable for collecting code statements
    String code = "package entities;\n\n",
    code += "import java.io.Serializable;\n\n";
    code += "public class " + cl.getName() + "implements Serializable{\n";

    // generate Attributes:
    Iterator<Attribute> attIter = cl.getAtts(); ... code += ...

    // generate Methods:
    Iterator<Operation> opIter = cl.getOps(); ... code += ...

    code += "}";

    // PHASE 3: print code to file
    try {
        FileOutputStream fos = new FileOutputStream(cl.getName() +".java");
        fos.write(code.getBytes());
        fos.close();
    } catch (Exception e) {.}
}
```

The advantages of this approach are that no additional programming skills are needed. It is sufficient to know the programming language chosen to develop the generator and to have familiarity with the model API. Furthermore, no additional tools are needed, neither for the design time nor for the runtime. However, following such an approach has also several drawbacks.

- **Intermingled static/dynamic code:** There is no separation of *static code*, i.e., code that is generated in exactly the same way for every model element, e.g., the package definition, the imports, etc., and *dynamic code* which is derived from model information, e.g., class name, variable name.

- **Non-graspable output structure:** The structure of the output is not easily graspable in the code generator specification. The problem is that the produced code is embedded into the producing code. Thus, the control structure of the code generator is explicit, but not the output format. This problem is also manifested in other GPL-based generator approaches such as in Java Servlets[1] for producing HTML code by statements embedded in the producing code.

- **Missing declarative query language:** There is no declarative query language for accessing model information available. Thus, many iterators, loops, and conditions as well as type

---

[1]http://www.oracle.com/technetwork/java/index-jsp-135475.html

casts unnecessarily lead to a huge amount of code. Also, be aware that knowledge of the generated model API is required. For example, to access features of model elements, getter methods have to be used instead of querying the feature values just by using the feature names defined in the metamodels.

- **Missing reusable base functionality:** Code has to be developed for reading input models and persisting output code again and again for each code generator.

To eliminate the mentioned disadvantages, DSLs have been developed for generating text from models. This has also led to an OMG standard called *MOF Model to Text Transformation Language*[2] (MOFM2T). In the following section, we show how the Java-based code generator may be re-implemented in a dedicated M2T transformation language and discuss the benefits of such an approach.

9.3   CODE GENERATION THROUGH M2T TRANSFORMATION LANGUAGES

In this section, we present how M2T transformation languages ease the development of code generators, give an overview of existing protagonists, and show how to use one of them to implement a code generator for our running example.

9.3.1   BENEFITS OF M2T TRANSFORMATION LANGUAGES

In the previous section, a Java-based code generator has been presented to show how the necessary features have to be implemented in GPLs. M2T transformation languages aim at improving the code generator development by tackling the aforestated drawbacks of GPL-based code generators.

Separated static/dynamic code.   M2T transformation languages separate static and dynamic code by using a *template*-based approach to develop M2T transformations. A template[3] can be seen as a kind of blueprint which defines static text elements shared by all artifacts as well as dynamic parts which have to be filled with information specific to each particular case. Therefore, a template contains simple text fragments for the static part and so-called *meta-markers* for the dynamic part. Meta-markers are placeholders and have to be interpreted by a *template engine* which processes the templates and queries additional data sources to produce the dynamic parts. Obviously, in M2T transformations, the additional data sources are models. Figure 9.4 summarizes the main idea of template-based text generation at a glance.

Explicit output structure.   Using templates allows us to explicitly represent the structure of output text within the template. This is achieved by embedding the code for producing the dynamic

---

[2]http://www.omg.org/spec/MOFM2T
[3]Please note that templates for code generation are different from templates in programming languages, e.g., in C++, which allow for datatype independent programming.

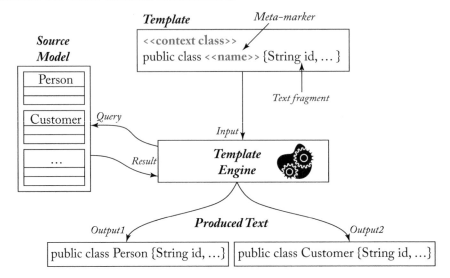

Figure 9.4: Templates, template engines, and source models to produce text.

parts of the output in the text representing the static part—exactly the inverse of the previous Java-based code generator. Here the same mechanism applies as for Java Server Pages[4] (JSPs), which allow us to explicitly represent the structure of the HTML pages and embed Java code—in contrast to Java Servlets. Having the structure of the output explicitly represented in the templates leads to more readable and understandable code generation specifications than just using String variables to store output text. Templates also ease the development of code generators. For example, a template may be developed by just adding example code to a template and the developer substitutes the dynamic code parts with meta-markers. By this: (i) the abstraction process from a concrete code example to a template specification is straightforward and (ii) the template has a similar structure and format as the code to be produced, which allows us to trace the effects of templates to the code level.

Declarative query language.   Within the meta-markers, we need to access the information stored in the models. As presented before, OCL is the choice to do this job in most M2M transformation languages. Thus, current M2T transformation languages also allow us to use OCL (or a dialect of OCL) for specifying meta-markers. Other template languages not specifically tailored to models but supporting any kind of sources employ standard programming languages such as Java for specifying meta-markers.

Reusable base functionality.   Current M2T transformation languages come with tool support which allow us to directly read in models and to serialize text into files by just defining configura-

---

[4]http://java.sun.com/products/jsp/reference/api/index.html

tion files. Thus, no tedious redefinition of model loading and text serializing has to be developed manually.

9.3.2 TEMPLATE-BASED TRANSFORMATION LANGUAGES: AN OVERVIEW

Different template-based languages exist which may be employed to generate text from models.

- **XSLT:** The XMI serializations of the models may be processed with XSLT,[5] which is the W3C standard for transforming XML documents into arbitrary text documents. However, in this case, the code generation scripts have to be implemented based on the XMI serialization which requires some additional knowledge on how models are actually encoded as XML files. Thus, approaches directly operating on the model level are more favorable.

- **JET:** The Java Emitter Template (JET) project[6] was one of the first approaches for developing code generation for EMF-based models. But JET is not limited to EMF-based models. In general, with JET, every Java-based object is transformable to text. JET provides a JSP-like syntax tailored to writing templates for M2T transformations. As for JSP, arbitrary Java expressions may be embedded in JET templates. Furthermore, JET templates are transformed to pure Java code for execution purposes. However, there is no dedicated query language for models available in JET.

- **Xtend:** Xtend[7] is a modern programming language which is mainly based on Java but offers several additional language features. For instance, it provides dedicated support for code generation in the form of template expressions. Furthermore, it supports functional programming, which is beneficial for querying models (especially many iterator-based operations from OCL are available out-of-the-box).

- **MOFScript:**[8] This project provides another M2T transformation language providing similar features such as Xtend. MOFScript has been developed as a candidate proposal in the OMG standardization effort providing a standardized language for M2T transformations. MOFScript is available as an Eclipse plug-in and supports EMF-based models.

- **Acceleo:**[9] The aim of this project is to provide a pragmatic version of the M2T transformation standard of the OMG for EMF-based models. The language provides full OCL support for querying models and mature tool support, which has proven useful in industry.

---

[5]http://www.w3.org/TR/xslt20/
[6]http://www.eclipse.org/modeling/m2t/?project=jet#jet
[7]http://www.eclipse.org/xtend
[8]http://www.eclipse.org/gmt/mofscript
[9]http://www.acceleo.org/

9.3.3   ACCELEO: AN IMPLEMENTATION OF THE M2T TRANSFORMATION STANDARD

Acceleo is selected as a protagonist to demonstrate M2T transformation languages in the following, because of its practical relevance and mature tool support. Please note that the language features of Acceleo are mostly supported by other M2T transformation languages such as Xtend or MOFScript as well.

Acceleo offers a template-based language for defining code generation templates. The language comes with a powerful API supporting OCL as well as additional operations helpful for working with text-based documents in general, e.g., advanced functions for manipulating strings. Acceleo is shipped with powerful tooling such as an editor with syntax highlighting, error detection, code completion, refactoring, debugger, profiler, and a traceability API which allows us to trace model elements to the generated code and vice versa.

Before templates can be defined in Acceleo, a module has to be created acting as a container for templates. The module also imports the metamodel definition for which the templates are defined. This makes the template aware of the metamodel classes that can now be used as types in the template. A template in Acceleo is always defined for a particular metamodel class. In addition to the model element type, a pre-condition can be defined which is comparable to a filter condition in ATL, e.g., to apply a template exclusively to model elements which are required to have a specific type as well as specific values.

The Acceleo template language offers several meta-markers which are called tags in Acceleo and which are also common in other available M2T transformation languages:

- **Files:** To generate code, files have to be opened, filled, and closed like we have seen before for the Java-based code generator. In Acceleo, there is a special `file` tag which is used to print the content created between the start and the end of the `file` tag for a given file. The path and the file name are both defined by an attribute of the tag.

- **Control Structures:** There are tags for defining control structures such as loops (`for` tag) for iterating over collections of elements, e.g., especially useful for working with multi-valued references obtained when navigations result in a collection of elements, and conditional branches (`if` tag).

- **Queries:** OCL queries can be defined (`query` tag), similar to helpers in ATL. These queries can be called throughout the whole template and are used to factor out recurring code.

- **Expressions:** There are general expressions for computing values in order to produce the dynamic parts of the output text. Expressions are also used to call other templates to include the code generated by the called templates in the code produced by the caller template. Calling other templates can be compared to method calls in Java.

- **Protected Areas:** An important feature of M2T languages is to support projects where only partial code generation is possible. In particular, special support is needed to protect manually added code from file modifications in subsequent code generator runs. For this task, a special concept named protected areas has proven useful, which is supported by Acceleo via the protected tag. Protected areas are used to mark sections in the generated code that shall not be overridden again by subsequent generator runs. These sections typically contain manually written code.

Code generation example.   The following listing shows an Acceleo template equivalent to the previously presented Java-based code generator (cf. Listing 9.1, p. 146). In the first line, the module called *generateJavaClass* is defined which includes the import of the sMVCML metamodel. The module itself is structured into one main template (cf. template *javaClass*) targeting the generation of code for classes sMVCML classes. This template delegates the code generation for attributes and operations to additional specific templates.

In the listing, several different tags are used. The `Query` tag is used to produce the signature names for getter and setter methods necessary for accessing/modifying attributes as well as for computing a default return statement based on the return type for ensuring the generation of compilable code. The `file` tag is used for opening and closing the file in which the code for the Java class is printed. `For` tags are used to iterate over the attributes and operations of a processed class for calling the specific templates.

Expressions are used several times. For instance, `[cl.name/]` prints the name of the class to the text stream. Other expressions are calling templates, e.g., `[javaAttribute(att)/]` is used to call the template for producing code for the attributes and `[att.getter()/]` is used to call the query to produce the name of the getter methods.

Listing 9.2: Acceleo-based Code Generation

```
[module generateJavaClass('http://smvcml/1.0')]

[query public getter(att : Attribute) : String = 'get'+att.name.toUpperFirst() /]

[query public returnStatement(type: String) : String = if type = 'Boolean'
    then 'return true;' else '...' endif  /]

[template public javaClass(aClass : Class)]

[file (aClass.name.toUpperFirst()+'.java', false, 'UTF-8')]
package entities;

import java.io.Serializable;

public class [aClass.name/] implements Serializable {

[for (att : Attribute | aClass.atts) separator ('\n')]
[javaAttribute(att)/]
[/for]

[for (op : Operation | aClass.ops) separator ('\n')]
[javaMethod(op)/]
[/for]

}
```

```
[/file]
[/template]

[template public javaAttribute(att : Attribute)]
  private [att.type/] [att.name/];

  public [att.type/] [att.getter()/]() {
    return [att.name/];
  }
  ...
[/template]

[template public javaMethod(op : Operation)]
  public [op.type/] [op.name/]() {
    // [protected (op.name)]
    // Fill in the operation implementation here!
    [returnStatement(op.type)/]
    // [/protected]
  }
[/template]
```

In the javaMethod template, a protected area is used to define the space for including the implementation of the operations. The produced output for this template is shown in Listing 9.3. Please note that in the first generator run, the protected area is created including only the standard comment and a default return value as given in the template as a placeholder for the actual method implementation. In all subsequent generator runs, this space is not changed again, which means that the user can implement the method and the manually added code is not lost in later generator runs.

Listing 9.3: Effect of Protected Areas

```
public boolean checkAvailability(){
    // Start of user code checkAvailability
    // Fill in the operation implementation here!
    return true;
    // End of user code
}
```

9.4   MASTERING CODE GENERATION

In order to deal with the complexity when implementing model-based code generators, several advanced techniques can be utilized which are briefly described in the following.

Abstracting templates. One important issue is the Turing test for code generation (cf. Chapter 3). To ensure that the generated code is accepted by developers, which is especially important when only a partial code generation can be applied, abstracting code generation templates from concrete reference examples is a proven approach. Acceleo offers the possibility to refactor a concrete reference example to a template. Several refactoring operations are available to substitute example code with tags. For instance, a concrete name of a Java class may be selected and refactored to an expression which queries the name value from a model element. By using this tool support, a template can be systematically extracted from example code, which ensures that developers will be familiar with the produced output of the templates when completed.

Reusing templates. Current M2T transformation languages provide several features to structure and reuse transformation logic. For example, templates may be structured into an inheritance tree which allows us to have polymorphic template calls, i.e., a call to a general template is always dispatched to the most specific template based on the model element type. This is particularly useful because in metamodels inheritance between classes is heavily used to reuse features. Thus, templates may be reused and extended for subclasses which have been defined before for superclasses to avoid duplicated code. Further features which are supported by M2T languages are aspects for defining and weaving cross-cutting concerns to a code generator or the refinement of complete transformation modules by overriding templates instead of extending them by inheritance.

Step-by-step generation. Dividing a code generation process into several steps by chaining transformations is useful in particular when there is still a big gap between the model level and the code level. This leads to verbose code generation templates which also hide the structure of the final output. Thus, applying the divide-and-conquer principle for developing code generations by using again the concept of transformation chains (cf. Section 8.4) having a mixture of M2M and M2T transformation steps results in more readable, reusable, and maintainable components. For example, assume a code generator for nested state machines has to be developed. If a code generator for flat state machines already exists, one could just provide an M2M transformation from nested state machines to flat state machines and reuse the existing code generator.

Separating transformation logic from text. Complex computations such as queries on the input model or complex computations of Strings should be refactored as much as possible into queries or libraries to be imported in the templates when needed. By this, the structure of the final output becomes more visible in the templates and previous defined code becomes reusable also in other templates or code generators.

Mastering code layout. The code layout is determined by the template layout. For example, it can be challenging to produce exactly the desired layout when several loops and conditions are comprised in the template. Some approaches such as Xtend allow us to have a code beautifier as a post processing unit for the code generator output. Besides employing code beautifier in a post-processing step, to have more control on the actual layout of the output, there are several language features in the M2T languages such as operators for ignoring line breaks in the templates, for separating elements by special signs produced in loops, etc.

Be aware of model/code synchronizing problems. Although the concept of protected areas is very helpful when following a partial code generation strategy, there may be cases where manually added code cannot be placed at the right position in the newly produced code of the next generator run. For example, assume the class *Tutorial* in our example is renamed to *Talk*. This means, in the next generator run, a new Java class is generated, but the code manually added to the (now old) Java class *Tutorial* may be lost. Especially when we use the name of the class concatenated with the name of the method as protected area ID, then rename refactorings on the class and on the method on the model level, this would make the synchronization between model and code impossible. If such kinds of refactorings are used, we end up having a new Java file with a "new" operation, and the implementation code for this operation has to be copied from the old Java file, added to the new class, and adapted manually to the new names. Some M2T transformation frameworks such as Acceleo at least report if some manually added code cannot be merged with the newly created code. In such cases you find a log file produced by the code generator summarizing the lines of code which have not been integrated in the new code version.

This trivial example shows quite well that one has to be careful when refactorings are applied on the model and a partial code generation has been already used. One way of dealing with such kinds of refactorings may be to execute refactorings on the model level and the corresponding refactorings on the code level, and run the code generator after the propagation of the refactorings. Another approach may be to just use the internal IDs of the model elements (which are immutable and automatically assigned by the modeling tools) for defining the IDs of the protected areas. While in the former approach more effort has to be spent to build such a model/code synchronization support, in the latter no additional effort is needed. However, the former approach may be powerful enough to adapt the previously manually added code automatically to the new model version, whereas this is not possible by the latter.

9.5   EXCURSUS: CODE GENERATION THROUGH M2M TRANSFORMATIONS AND TCS

If the target language of an M2T transformation is already supported by a metamodel and a TCS, which allows us to parse/serialize models from/to text, the M2T transformation may be realized by using an M2M transformation instead. Figure 9.5 illustrates the resulting transformation process if such a methodology is followed. First a transformation from the metamodel of the modeling language to the metamodel of the text-based language is performed, and subsequently, the produced target model is serialized into a text based on the mapping between the abstract syntax and the TCS.

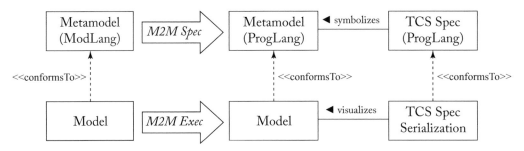

Figure 9.5: Using TCS for code generation.

This approach seems beneficial at a first sight: (i) reuse of the textual concrete syntax definition ensures producing valid textual artifacts; (ii) having an explicit metamodel for the text-based language allows us to better reason about the language concepts and the mappings from the modeling language to the text-based language; (iii) transformation definitions may be validated based on the metamodels; (iv) output models representing the text-based artifacts can be validated w.r.t. the target metamodel; and (v) changes in the concrete syntax of the target language are independent of the transformation between the two languages, thus, no additional maintenance of code generators is necessary.

However, there are also current limitations of this approach. Most of the currently available metamodels for programming languages lack TCS definitions, or even more aggravated, no metamodels exist for the languages or they are not completely specified. Another drawback may be that the text-based languages are only known by their TCS. Thus, reasoning on the language concepts on the abstract syntax level may take some learning time, which is not necessary when working with code generation templates. The abstraction from reference code to transformation definitions is not supported in this approach by just copying the code and substituting some parts of it with meta-markers as is possible with M2T transformations. Furthermore, some features needed for partial code generation scenarios are not covered, such as protected areas. Finally, for a code generator, only a restricted part of the target language may be needed, which is much easier to implement with an M2T transformation language without reasoning on the complete language definition of the target language.

To summarize, although there are some benefits in reducing an M2T transformation to an M2M transformation, from a pragmatic viewpoint, it seems easier to start with building a code generator by employing M2T transformation languages—especially when no metamodel and TCS definitions for the target language exists. Thus, the M2M/TCS approach for code generation should be only considered when the target language is already supported by a metamodel and a TCS and a full code generation approach is followed.

CHAPTER 10

Managing Models

Creating models, metamodels, and transformations is only the beginning of an MDSE project, because when *everything is a model* (analogous to the popular statement: everything is an object), then *no model is an island* (analogous to no object is an island). This means we have many different kinds of models with many different relationships between them. And all these models and relationships must be properly managed during the project lifecycle. For instance, assume the consequences for a model if its metamodel is changed—the basic *conformsTo* relationship between them may no longer hold after the change and will need to be fixed.

This chapter presents different techniques for model management, including interchanging, persisting, comparing, versioning, and evolving models and their relationships. Furthermore, we also present tools and techniques for dealing with modeling-in-the-large, model quality, and collaborative modeling.

10.1 MODEL INTERCHANGE

Given the variety of modeling tools available we need a mechanism that allows the interchange of models between them. Paraphrasing the Java slogan "*write once, run everywhere,*" modelers would like to have a "*model once, open everywhere*" principle.

This is a major concern for developers, who are limited by the poor import/export features of model editors and often need to struggle finding by themselves the mappings and bridges between the formats of two different tools that need to interoperate. This hampers the possibility of effectively exploiting different tools in the development process. Clearly, this is an important drawback for the developers who cannot simply choose the best modeling tool for the job at hand. For instance, in a perfect world a developer could design the models with a modeling tool MT1 (e.g., because the editor of MT1 is more usable and allows higher productivity) and then immediately move to another tool MT2 for the verification phase (e.g., because MT2 provides more precise or extensive rules), and to tool MT3 for model execution.

Being a recognized critical problem, interoperability is being addressed by standardization bodies. They are well aware that the supposed role of facilitating communication and information interchange that (standard) models should cover is not a reality yet. Even in the case of well-established standards like UML or BPMN, their abstract definition is still not enough for leading modeling tools from different vendors to share their models in a seamless way. Concretely, each tool stores and manages the models with its own internal format or its own "dialect," even if a standard format is adopted. To improve interoperability between modeling tools, standardization bodies have defined specific model interchange languages.

The best-known model interchange language is XMI[1] (XML Metadata Interchange), a standard adopted by OMG for serializing and exchanging UML and MOF models. Unfortunately, even if this has been devised as an interchange format, different tools still adopt it with different flavors. Even the simple UML example model in Figure 10.1 is stored as two completely different XMI files depending on the tool used to model it.

Figure 10.1: UML example model.

This is the (simplified) XMI file generated for the UML example model when using the MDT/UML2 Eclipse project[2]:

```
<packagedElement xmi:type="uml:Class" xmi:id="_c001" name="Employee">
    <ownedAttribute xmi:id="_a001" name="name"/>
</packagedElement>
<packagedElement xmi:type="uml:PrimitiveType" xmi:id="_t001" name="String"/>
<packagedElement xmi:type="uml:Class" xmi:id="_c002" name="Department">
    <ownedAttribute xmi:id="_a002" name="name" type="_t001"/>
</packagedElement>
<packagedElement xmi:type="uml:Association" xmi:id="_as001" name="WorksIn"
    memberEnd="_e001 _e002">
    <ownedEnd xmi:id="_e001" type="_c002" association="_as001"/>
    <ownedEnd xmi:id="_e002" name="" type="_c001" association="_as001">
      <upperValue xmi:type="uml:LiteralUnlimitedNatural" xmi:id="un001" value="*"/>
    </ownedEnd>
</packagedElement>
```

and this is the (simplified) one produced by the open source ArgoUML tool, which is not even conformant to the XMI syntax:

```
<UML:Class xmi.id = '_c001'
    name = 'Employee' visibility = 'public' isSpecification = 'false' isRoot = 'false'
    isLeaf = 'false' isAbstract = 'false' isActive = 'false'>
    <UML:Classifier.feature>
        <UML:Attribute xmi.id = '_a001'
            name = 'name' visibility = 'public' isSpecification = 'false'
            ownerScope = 'instance' changeability = 'changeable' targetScope = 'instance'>
            <UML:StructuralFeature.multiplicity>
                <UML:Multiplicity xmi.id = '_m001'>
                    <UML:Multiplicity.range>
                        <UML:MultiplicityRange xmi.id = '_mr001'
                            lower = '1' upper = '1'/>
                    </UML:Multiplicity.range>
                </UML:Multiplicity>
            </UML:StructuralFeature.multiplicity>
            <UML:StructuralFeature.type>
                <UML:Class xmi.idref = '_st001'/>
```

---

[1]http://www.omg.org/spec/XMI/2.4.1
[2]http://wiki.eclipse.org/MDT-UML2

```
            </UML:StructuralFeature.type>
         </UML:Attribute>
      </UML:Classifier.feature>
</UML:Class>
<UML:Class xmi.id = '_c002'
    name = 'Department' visibility = 'public' isSpecification = 'false' isRoot = 'false'
    isLeaf = 'false' isAbstract = 'false' isActive = 'false'>
    ...  as before for the name attribute .....
</UML:Class>
<UML:Association xmi.id = '_as001'
    name = 'WorksIn' isSpecification = 'false' isRoot = 'false' isLeaf = 'false'
    isAbstract = 'false'>
    <UML:Association.connection>
       <UML:AssociationEnd xmi.id = '_ae001'
          visibility = 'public' isSpecification = 'false' isNavigable = 'true'
          ordering = 'unordered' aggregation = 'none' targetScope = 'instance'
          changeability = 'changeable'>
          <UML:AssociationEnd.multiplicity>
             <UML:Multiplicity xmi.id = '_m001'>
                <UML:Multiplicity.range>
                   <UML:MultiplicityRange xmi.id = '_mr001'
                      lower = '1' upper = '-1'/>
                </UML:Multiplicity.range>
             </UML:Multiplicity>
          </UML:AssociationEnd.multiplicity>
          <UML:AssociationEnd.participant>
             <UML:Class xmi.idref = '_c001'/>
          </UML:AssociationEnd.participant>
       </UML:AssociationEnd>
       ... same for the second association end ...
    </UML:Association.connection>
</UML:Association>
```

To overcome this situation, several tools are adopting as a de-facto standard the XMI implementation used by the Eclipse UML2 plugin. Since many UML tools are built on top of this component, the other UML tools also are forced to offer some import/export capabilities to ensure their interoperability to avoid losing potential customers.

Fortunately, the OMG has realized the importance of the problem and has now created the Model Interchange Working Group[3] (MIWG) to "enable the public at large to assess model interchange capability of the modeling tools by comparing the vendor XMI exports to the expected reference XMI file for each test case." So far, the working group includes Artisan Studio, RSx, IBM Rhapsody, MagicDraw, Modelio, and Enterprise Architect. The current test suite comprises a few dozen UML and/or SysML test models covering all major elements of the specification. This situation is leading to the definition of a canonical version of XMI (called Canonical XMI), a strict subset of XMI that forces the adopters to follow stricter constraints.

Note that an XMI file only contains information about the model elements. Graphical information (positions, colors, layout, fonts, etc.) is not part of the XMI file and thus it is lost in the interchange process, quite disappointing since designers usually invest a lot of effort in improving the presentation of models (i.e., moving classes around to minimize line crossing) as key factors in their readability and understandability. Some years ago, the OMG created a Diagram Interchange format to standardize the exchange of graphical model information, but it did not prove practical and was mostly unimplemented. The new Diagram Definition (DD)

---

[3]http://www.omgwiki.org/model-interchange

OMG standard[4] tries to fix this, facilitating the definition of the mappings between the model elements and their graphical information.

A similar situation can be found in many other modeling languages too. For instance, BPMN has an interchange counterpart in the XPDL language (XML Process Definition Language). However, this interchange language suffers from the same problems as the XMI. It is very difficult to actually move BPMN projects from one modeling environment to another, although BPMN 2.0 provides an XMI interchange format and also another non-XMI interchange format defined in the standard.

In this section we focused on interoperability problems at the syntactic level but, obviously, things get much more complicated if the tools involved follow different semantics, use proprietary extensions, or cover only a subset of the metamodel the models to interchange conform to. In such cases model transformations may be employed to resolve the heterogeneities between tools.

10.2   MODEL PERSISTENCE

Simple files are the most common storage format for models. Typically, tools use an XML-based serialization format (the above-mentioned XMI schema or a proprietary one) to store all the model information. This is, for instance, the default behavior of EMF-based tools in Eclipse. Each model is saved in a different file though cross references to external files are allowed.

This solution works well for most projects but it does present scalability issues when dealing with very large models (e.g., models with millions of instances) as, for instance, the ones generated as part of model-driven reverse engineering scenarios. For very large models, loading the full model in memory may not be a valid option, and random access strategies are required such as lazy loading. We highlight some tools that can be used to overcome this problem.

- **CDO:** The CDO (Connected Data Objects) Model Repository[5] was created with this goal in mind. CDO is a run-time persistence framework optimized for scalable query and transactional support for large object graphs. As back-end, CDO supports different strategies (object, NoSQL, and relational databases) though its main focus is on the latter. For relational databases, CDO relies on Teneo,[6] a Model-Relational mapping and runtime database persistence solution for EMF that can also be used as a stand-alone tool.

- **NeoEMF:** A NoSQL multi-backend framework for persisting and querying large models. It recognizes that there is no perfect backend for model storage and therefore facilitates the addition of custom backends for specific needs. By default, it includes a graph-based backend (compatible with Blueprints databases, especially Neo4J[7]) and a MapDB backend.

---

[4]http://www.omg.org/spec/DD
[5]http://wiki.eclipse.org/CDO
[6]http://wiki.eclipse.org/Teneo
[7]http://neo4j.com

Other NoSQL prototype tools are **Morsa**[8] and **MongoEMF**,[9] both using MongoDB as backend.

- **EMF Fragments:** EMF Fragments[10] differs from the previous approaches by focusing on the persistence of model chunks (fragments) instead of individual objects as key strategy. Each chunk is mapped to a URI and can be stored on a wide range of distributed data stores including key-value approaches.

- **IncQuery:** A high performance graph search. Instead of focusing on the storage aspect, IncQuery[11] focuses on the query one by providing incremental search capabilities for loaded models. In particular, runtime performance is achieved by adapting incremental graph pattern-matching techniques based on the Rete algorithm.

Using the cloud as a storage infrastructure for large models is also a promising future option to face these scalability issues as we will see later on in Section 10.8.

10.3   MODEL COMPARISON

The detection of differences between models is a crucial operation for model management in general. Understanding the differences between two models is as necessary as understanding the differences between two text-based artifacts. Apart from human understanding, differences between models are a necessary input for automating model management activities such as metamodel/model co-evolution (cf. Section 10.5) or model versioning (cf. Section 10.4). These tasks are based on so-called *difference models*, which—as their name suggests—are models stating the differences between two models. Producing the difference models is the job of model comparison tools which take two models as input and compute the difference model as output.

Model comparison is implemented as a two-phase process [12]. First, the two models have to be matched, i.e., the corresponding elements in the two versions are marked, and second, based on the match, differences are derived.

Phase 1: Model matching. Matching two model elements means to reason about their *identity*. Model elements with equal identities are considered a match. There are different matching strategies which employ different characteristics of the model elements to establish their identity. Kolovos et al. [46] distinguish between static identity-based matching, signature-based matching, similarity-based matching, and language-specific matching. Static identity-based matching relies on artificial unique identifiers attached to each model element, whereas signature-based matching compares model elements based on a computed combination of their feature values. Approaches in these two categories treat the problem of model matching dually,

---

[8]http://modelum.es/trac/morsa
[9]https://github.com/BryanHunt/mongo-emf/wiki
[10]https://github.com/markus1978/emf-fragments
[11]https://www.eclipse.org/incquery

i.e., two model elements match or do not match. In contrast, similarity-based matching computes a similarity measure between two model elements based on name equivalences or structural equivalences such as incoming/outgoing links. Finally, language-specific matching requires us to specify dedicated match rules which allows us to consider the underlying semantics of the modeling language.

   Phase 2: Model differencing. Differencing algorithms perform a fine-grained comparison of two model elements that correspond to each other, i.e., they have been matched in phase 1. If two corresponding model elements differ in some way, a description of the difference is created and added to the difference model. Existing model comparison approaches are capable of detecting differences that correspond to the application of *atomic operations*, i.e., add, delete, move, and update. For example, if a model element has no corresponding model element on the opposite side (i.e., there is no match for this element), an element insertion or deletion is noted—here it depends on which model is considered to be the initial model and which the revised one.

For EMF models several model comparison tools and frameworks exist.

- **EMF Compare**[12] is an Eclipse project that provides generic comparison facilities for any kind of EMF model (tailoring the comparison algorithm to types of EMF models is also a possibility). The calculated differences between two models are visualized and can be exported in a model patch with similar functionality to what is known from program patches. Comparison support for graphical diagrams is also under development.

- **SiDiff**[13] is a meta model-independent approach for model comparison. It is primarily based on the notion of similarity between model elements, but covers other approaches like identity-based or signature-based model comparison as well. The main advantage of SiDiff is that it offers a highly configurable environment and is therefore easily adaptable to any modeling language where models can be represented in a graph-like structure.

- **Epsilon Comparison Language**[14] (ECL) allows the implementation of specialized comparison algorithms by using a dedicated DSL. Thus, higher-level changes, going beyond atomic changes, such as refactorings, may be also detected. ECL can compare two models conforming to two different metamodels. For example, it would be possible to develop comparison support with ECL for comparing UML class diagrams with Entity Relationship diagrams.

---

[12]http://www.eclipse.org/emf/compare
[13]http://pi.informatik.uni-siegen.de/sidiff
[14]http://www.eclipse.org/epsilon/doc/ecl

10.4   MODEL VERSIONING

Version Control Systems (VCS) are a very important part of the software development infrastructure which: (i) store the evolution history of software artifacts; (ii) support multiple developers working in parallel; and (iii) manage different development branches [13]. For all of these tasks, changes performed on the artifacts under version control have to be tracked. For the second and third task it is additionally necessary to detect conflicts between concurrently evolved versions of one artifact and to resolve such conflicts in order to obtain a consolidated version. When merging two concurrently modified versions of a software artifact, conflicts might inevitably occur. VCS help to detect, manage, and resolve these conflicts.

Clearly, programmers could not live without VCSs like Subversion (SVN) or Git. However, modelers have been forced to live like that until recently. Things are changing and model versioning techniques have started to appear [12].

Traditional text-based versioning systems treat models just as plain text files and, as a consequence, neglect the graph-based nature of models. A simple example showing the impedance mismatch between models and their text-based representations [52] in the context of model versioning is the following. Assume that we have a basic state machine diagram containing two states and one transition (cf. *V0* in Figure 10.2). Now this model is concurrently edited by two modelers, which leads to two different versions of this model (cf. *V1'* and *V1"*). In the left version, only the second state remains which indicates that the first state and the transition have been deleted. In the right version, all elements of the initial diagram remain and in addition a new transition has been introduced. A text-based versioning system working at the XMI serialization level would produce an automatically merged version *V1* for this case where state B and a dangling transition pointing to nowhere exist.

The reason for this merge result produced by a text-based VCS is that on the text level the VCS only reasons about lines which have been changed or deleted. In our example, the text-based comparison algorithm may detect that the first line and the last line have been deleted by comparing the left version (*V1'*) with the original version. Analogously, when comparing the right version with the original version, the addition of a new line is detected. However, no support is given when interpreting models only as lines of text for detecting that the referenced state A of the added transition has been concurrently deleted, which should definitely lead to a conflict.

This example shows that it is possible to version the text-based serialization of models. However, several issues may remain undetected. Thus, dedicated versioning support for models for detecting possible conflicts are required. Fortunately, some interesting model versioning tools are already available (some of them built on top of the model comparison tools presented in the previous section).

- EMFStore project[15] is the official Eclipse project devoted to provide a model repository able to keep track of the version history of the stored models. EMFStore follows the check-

---

[15]http://www.eclipse.org/emfstore

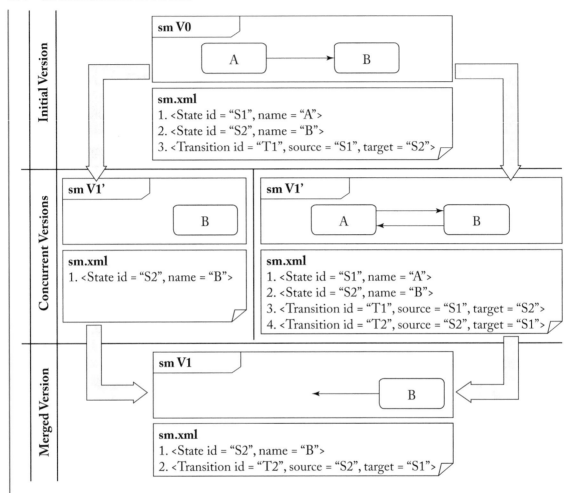

Figure 10.2: Text-based versioning of state machines.

out/update/commit interaction paradigm known from SVN and offers an interactive model merging interface to resolve conflicts when two users apply overlapping changes on model elements. A model migration feature (to update models after applying changes to the metamodels they conform) is also integrated.

• The AMOR (Adaptable Model Versioning) project[16] provides methods and techniques to leverage version control for EMF-based models. Several conflict detection, visualization, and resolution possibilities are provided. In particular, merge conflicts are visualized in the graphical syntax of models using model annotation techniques.

---

[16]http://www.modelversioning.org

- The CDO model repository (cf. Section 10.2) includes limited branching and merging support for models on the database level.

- Epsilon Merging Language[17] is a rule-based language for merging homogeneous or heterogeneous models. The correspondences between the models to be merged can be generated using the related Epsilon Comparison Language (cf. Section 10.3).

Note that all these tools only version the model information but typically not its graphical representation (e.g., the diagram layout). One possibility would be to version the diagram information as well, which is also represented as a model in many graphical model editors. Nevertheless, when considering the concurrent modification of a diagram by different modelers, it is still interesting to discuss what should be considered as a change or conflict in this context. For instance, should moving a class two inches to the right count as a change? Does this conflict with another "change" in which the class has moved a little bit down? Diagram versioning is still an open issue.

10.5 MODEL CO-EVOLUTION

Model versioning tools track the evolution of one modeling artifact during its life span. But this is just one side of the evolution problem. Since models are not isolated entities, but interrelated, a change in one of them may impact others. This means that if one model evolves, dependent models have to co-evolve in order to reestablish the relationships between them.

The most typical conflictive evolution scenario is the evolution of a modeling language through the modification of its metamodel [50] in order to correct errors, support new modeling features, refine others, etc. This metamodel evolution requires us to co-evolve the already existing model instances, which may no longer conform to the new metamodel version (e.g., if we change the type of an attribute we may need to update the values of all models for that same feature so that their value is consistent with the new type). Affected models have to be migrated to the new metamodel version.

Figure 10.3 illustrates the model migration problem at a glance. Full arrows are transformations; dashed arrows are representing the conformsTo relationships between models and metamodels. After evolution of a metamodel MM (cf. upper half of the figure) which can be seen as a transformation T_{MM} of the metamodel, the goal is to migrate models M (conforming to MM) to M' (conforming to MM') by creating a suitable migration transformation T_M (cf. lower half of the figure). But before models can be migrated to the new metamodel version, the changes between the previous and the new metamodel version have to be carefully determined, because they form the basis for adapting the models.

In this context it is important to categorize the consequences of metamodel changes on the models before evolving the metamodel [17].

---

[17]http://www.eclipse.org/epsilon/doc/eml

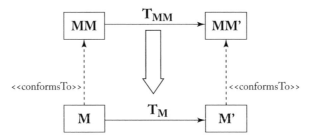

Figure 10.3: Evolution of metamodels and co-evolution of their corresponding models.

- **Non-breaking operations:** A change at the metamodel level requires no migrations of the instances, e.g., adding a new optional attribute to a metaclass.

- **Breaking and resolvable operations:** A change at the metamodel level is reflected on the instance level by an automatic migration, e.g., adding a mandatory attribute with a default value to a metaclass.

- **Breaking and unresolvable operations:** Some changes at the metamodel level need complex migrations which possibly require user input to introduce additional information to the models, e.g., adding a mandatory attribute for which no default value is available to a metaclass.

Depending on the complexity of the metamodel evolution, the migration transformation needed to adapt the models to the new metamodel version may be derived from a difference model with the list of changes done to the metamodel during its evolution. Typically, this can be achieved for breaking and resolvable operations. As an example, consider that a metaclass is deleted; a default migration strategy is to just delete its instance in all models too. This leads to models conforming again to the metamodel from a syntactical point of view, but whether this default migration is valid from a semantical point of view has still to be decided by the user. For instance, for some cases the instances may have to be casted to other metaclasses instead of just being deleted.

Several tools can assist in the migration process [59].

- **Edapt**[18] provides support for the evolution of Ecore metamodels and the automatic migration of models to the new metamodel version by providing standard migration strategies.

- **Epsilon Flock**[19] is a model migration language built atop the Epsilon family of languages, for automatically updating models in response to metamodel changes.

- Any **M2M transformation language** may be used to specify migration paths for models for dealing with co-evolution.

---

[18]http://wiki.eclipse.org/Edapt
[19]http://www.eclipse.org/epsilon/doc/flock

Other co-evolution scenarios   The model/metamodel co-evolution scenario is just one of the examples of the impact that a change on a modeling artifact can have on other related ones. As an example, changing a metamodel can also affect all transformations that read or write models of that type. After a change, these transformations may become invalid since, for instance, they may reference model elements that no longer exist. Also, OCL constraints may have to be reconciled when their metamodel evolves. And the same for concrete syntax specifications. Of course, all these co-evolution scenarios result from the fact that metamodels are in the center of the language engineering process, thus changes to them have to be propagated to all depending artifacts.

Besides co-evolution at the metamodeling level, also at the model-level several co-evolution scenarios exist. Just consider a multi-view modeling language such as UML. Assume an operation is deleted in a class diagram. If state machines are using this deleted operation as a trigger for transitions, the models are in an inconsistent state. Thus, also on the model-level, co-evolution is an important aspect which should be supported by inconsistency detection rules, e.g., defined in OCL, and co-evolution rules, e.g., defined with M2M transformations.

10.6   GLOBAL MODEL MANAGEMENT

So far we have seen many techniques and tools to manage different aspects of the models' lifecycle in a connected modeling ecosystem. The global problem of managing this ecosystem is known as Global Model Management (GMM). This is also the key challenge behind the international initiative on the "Globalization of Modeling Languages, a language workbench for heterogeneous modeling to support coordinated development."[20]

A key element to manage the ecosystem as a whole is keeping in one place (a model) all the information about the involved models and the relationships between them. This model is known as a *megamodel* [23]. Megamodels include not only the project artifacts (MDSE-based like models but possibly also non-MDSE artifacts like configuration files and others) but also their relationships and other relevant metadata (such as the actual type of an artifact, its identifier, location, and so on) required to manage them. In this sense, a megamodel can be viewed as a metadata repository where precise representations of models and the relationships between them are stored and made available to users for various purposes.

A megamodel is a normal model with the only particularity that its model elements are in fact other models. As a model, a megamodel must conform to a metamodel. An example of a concrete metamodel for megamodels has been implemented as part of the Eclipse AM3 (AtlanMod MegaModel Management) megamodeling framework.[21] A subset of this metamodel is shown in Figure 10.4. As you can see in the figure, the types of the elements in a megamodel can be models, metamodels, transformations and any other kind of modeling element relevant for an MDSE project.

---

[20]http://gemoc.org
[21]http://wiki.eclipse.org/AM3

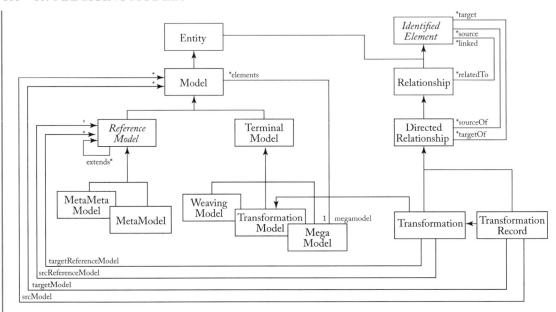

Figure 10.4: A metamodel for megamodels: the AM3 case.

An important benefit is that, as a model, a megamodel can be manipulated using the same tools available for other models. Nevertheless, specialized tools for model management also exist. In particular, we highlight *MoScript*,[22] a textual DSL and megamodel agnostic platform for accessing and manipulating modeling artifacts represented in a megamodel. MoScript allows the writing of queries that retrieve models from a repository, to inspect them, to invoke services on them (e.g., transformations), and the registration of newly produced models into the repository. MoScript scripts allow the description and automation of complex modeling tasks, involving several consecutive manipulations on a set of models. As such, the MoScript language can be used for modeling task and/or workflow automation and in general any batch model processing task you can imagine. The MoScript architecture includes an extensible metadata engine for resolving and accessing modeling artifacts as well as invoking services from different transformation tools.

As an example, suppose we want to regenerate our Java systems in .NET. Assuming that we already have a transformation java2dNet, we may want to apply this transformation to all Java models in our repository to generate all the corresponding .NET models. The following listing shows how MoScript could automatically and autonomously do that. The script first selects the set of models conforming to the Java metamodel from the repository. Then the transformation java2dNet is also retrieved from the repository and applied to each Java model by means of the

---

[22]http://wiki.eclipse.org/MoScript

applyTo operation. The result of this example is a collection of TransformationRecord elements, which contain the newly produced .Net models.

```
let javaModels : Collection(Model) =
  Model::allInstances()->select(m | m.conformsTo.kind = 'Java') in
let java2dNet : Transformation =
  Transformation::allInstances()->any(t | t.identifier = 'java2dNet') in
javaModels->collect(jm | java2dNet.applyTo(Sequence{jm}))
```

Behind the scenes, MoScript does a type check between the metamodels of the models to be transformed and checks the compatibility between the metamodels and the transformations by querying the megamodel. In the same way MoScript determines which transformation engine should execute the transformation. Finally, MoScript sends the models, metamodels, and transformations to the transformation engine for execution. The resulting models may be persisted in the model repository, inspected, or even transformed again with a chained MoScript script.

10.7   MODEL QUALITY

The change of perspective that MDSE brings to software engineering implies that the correctness of models (and model transformations, also a kind of model as discussed before) becomes a key factor in the quality of the final software product. For instance, when code is no longer written from scratch but synthesized from models (semi-)automatically, any defect in the model will be propagated into defects in the code.

Although the problem of ensuring software quality has attracted much attention and research, it is still considered to be a *grand challenge* for the software engineering community [39]. It is essential to provide a set of tools and methods that helps in the detection of defects at the model-level and smoothly integrates in existing MDSE-based tool-chains without an excessive overhead.

In general, current modeling tools only ensure the well-formedness of the models; that is, tools can check whether the model you are specifying is a valid instance of the metamodel the model conforms to. This verification can be done on-demand (the designer must explicitly trigger the verification and is responsible for correcting the errors) or automatically (the tool prevents adding new modeling elements that will break well-formedness). For instance, the UML2 metamodel implementation in Eclipse offers on demand validation of some UML well-formedness rules. This feature can be used by Eclipse-based UML modeling tools to offer a somehow "safe" modeling environment to their users.

Well-formedness is usually only considered for a single model. Consistency between different views of the system expressed using different but complementary models is not enforced. This makes sense at the early stages of the development process (since these models may be developed by independent teams of designers, e.g., the method calls defined in a UML sequence diagram modeled by one team cannot correspond to actual methods in the classes of the class diagram defined by a different team), but at some point in time it is important that these inconsistencies disappear.

10.7.1 VERIFYING MODELS

Well-formedness is just the tip of the iceberg when it comes to model quality. A model can be well-formed but still be incorrect. Even very simple models can turn out to be incorrect. Take a look at the model in Figure 10.5, representing the relationships between students and the courses they enroll in and those that they like.

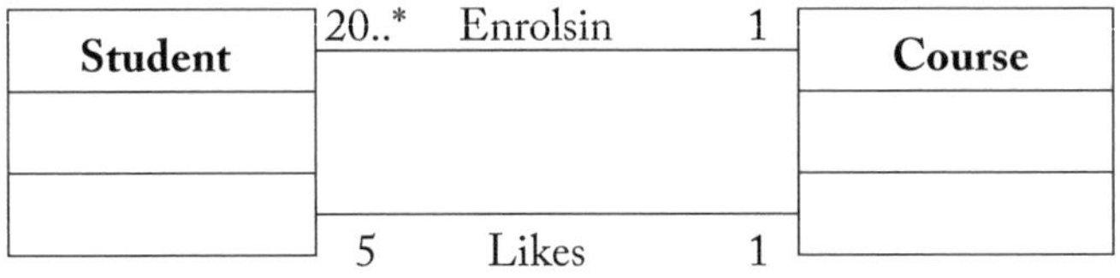

Figure 10.5: Well-formed but invalid model: unsatisfiability problem.

Even if the model is small and has no OCL constraints, it turns out that it is impossible to instantiate the model in a way that all cardinality constraints in the associations become satisfied at the same time (this is known as the *satisfiability* problem and many more advanced correctness properties, like redundancy, are defined based on this one). Therefore, the model is useless since users will never be able to use it to store any consistent information. The error in this model is that the *EnrollsIn* association forces the system to have at least 20 students per course while *Likes* states that the number of students per course must be exactly 5.

For this example, and given that the problem lies only the cardinality constraints, a verification tool could easily detect the errors and provide adequate feedback to the user, but unfortunately the problem of checking the correctness of an arbitrary model (including any combination of modeling constructs plus integrity constraints expressed in an expressive language like OCL) is a very complex one (in fact, it is not even decidable in general, meaning that it is not possible to create an algorithm that decides whether an arbitrary model is correct or not). This explains why so far we do not have any available verification tool that can be easily integrated with state-of-the-art modeling tools. Existing verification tools require designer interaction, restricting the possible elements to be used in the models, and deep knowledge of formal methods or extensive manual model annotations which seriously impair its usability in practice.

To avoid these problems, a new generation of verification tools follows a more pragmatic approach where, instead of trying to find a perfect answer, the goal is to give an answer (with a certain degree of confidence) in a reasonable amount of time. An example of such tools is EMFtoCSP,[23] a tool that checks UML and EMF models by transforming them transparently to a constraint satisfaction problem [70] that is then sent to a constraints solver to evaluate the satisfiability of the model.

The relevant correctness properties for each model depend on the type of the model. For static models (such as UML class diagrams) typical properties involve the satisfiability property

---

[23]https://github.com/SOM-Research/EMFtoCSP

mentioned before (i.e., it must be possible to create a valid instantiation of the model) and absence of redundancies. For dynamic models, useful properties involve the absence of deadlocks and infinite recursion as well as the applicability and executability of operations.

10.7.2 TESTING AND VALIDATING MODELS

Formal methods are not the only solution to evaluate the quality of models. Several testing approaches have been proposed to test models similarly to what software testing techniques do at the code level.

This is especially true for UML models annotated with OCL constraints. Tools like USE[24] can automatically create system states (snapshots of a running system) from a model. For each snapshot, OCL constraints are automatically checked and the results are given to the designer using graphical views. This simulation of the system allows designers to identify if the model is overconstrained (i.e., some valid situations in the domain are not allowed by the specification) or underconstrained (some invalid scenarios are evaluated as correct in the specification). Similarly, several tools offer simulation capabilities for behavior models like state machines where you can see the evolution of the state machines depending on the system state and input values.

Testing techniques are of special interest to check the behavior of model transformations. In this sense, several white-box and black-box testing techniques have been proposed [7] that, as for the testing of programs, try to maximize the coverage while minimizing the number of tests to generate. The goal of a test for a model transformation would be to detect cases in which the transformation generates an invalid target model from a valid source one.

10.7.3 REVIEWING MODELS

Besides the classical validation, verification, and testing approaches for models, further support is needed to ensure and improve the quality of models by human inspection. One classical approach for this is reviewing, as it has been already used for programs for a long time by informal walk-throughs or formal inspections. Of course, these techniques are also needed on the model level to ensure the quality.[25]

Currently, there are not many approaches and tools for performing model reviews. A pioneer on this is EGerrit,[26] an Eclipse plug-in that provides an integration of Gerrit in Eclipse. In addition, EGerrit is a review tool that provides also dedicated support for modeling projects by combining EMF Compare and Gerrit. For instance, models and model differences may be directly commented on and review tasks such as voting are supported. While EGerrit provides first tool support for performing model reviews, more research is needed in the future to understand which review processes are beneficial for modeling projects.

---

[24]http://useocl.sourceforge.net/w/index.php/Main_Page
[25]http://agilemodeling.com/essays/modelReviews.htm
[26]http://eclipse.org/egerrit

10.8   COLLABORATIVE MODELING

Modeling is by definition a team activity. Team members discuss and refine models until they reach a consensus. As long as all members are in the same physical location this does not pose a problem. However, in the current context where most teams are distributed in different locations, modeling tools must offer collaborative modeling features that support online editing of models (offline collaborations can be managed by the model comparison and versioning tools presented before).

Online collaborative modeling systems rely on a short transaction model, whereby a single, shared instance of the model is edited by multiple users in real time (i.e., all changes are propagated to all participants instantly). These systems currently lack conflict management, or only provide very lightweight mechanisms (such as voluntary locking). As a result, conflict resolution is usually achieved by explicit consensus among all online parties, which involves a certain communication overhead.

Although still limited in number, there is a growing number of projects that focus on collaborative modeling. As they mature, we expect to see them integrated in many "traditional" modeling tools.[27]

- SpacEclipse-CGMF[28] is an attempt to integrate online collaborative functionality in GMF-based graphical editors. It also integrates a model-based way to define both the domain of the graphical editor and the workspace configuration of the tool to be generated, i.e., the user can provide a model specifying the desired features of the collaboration environment to be generated.

- Dawn[29] is a subproject of CDO aimed at providing collaborative access for GMF diagrams. Dawn (will) provide(s) predefined implementation strategies for conflict visualization and conflict handling and other useful features for building collaborative GMF editors.

- Collaboro[30] complements other approaches by providing a more participative process where designers can argue and vote for the different change proposals when evolving a (meta)model. All discussions and agreements are stored and therefore, design decisions can always be traced back to their original motivation.

We can also regard these tools as first steps toward the use of the cloud as a future core infrastructure for offering modeling services. We refer to this evolution as MaaS (Modeling as a Service). Still in early research stages, MaaS could be a solution to scalability problems appearing when loading or manipulating very large models and facilitate the combination of various modeling services (possibly from different vendors) in a single MDSE project. As an example,

---

[27]Here we focus on projects under the umbrella of the Eclipse Modeling Framework. Several commercial tools offer proprietary solutions as well.

[28]http://chico.inf-cr.uclm.es/jgallardo/space.html

[29]http://wiki.eclipse.org/Dawn

[30]https://github.com/SOM-Research/collaboro

EMF-Rest[31] generates a RESTFul API for your models so that you can easily access and manage them via the web.

---

[31]http://emf-rest.com

CHAPTER 11

Summary

The increasing complexity of software artifacts demands a more thorough application of model-driven techniques, so as to raise the level of abstraction at which the work is performed. Therefore, MDSE is shifting and will be further changing the way most software will be developed in the near future. That is why all software companies need to grasp the main concepts behind MDSE in order to understand *when*, *where*, and *how* MDSE can be introduced in their daily practice to get the most out of it.

This is the purpose of this book. The different chapters aim to give a complete description of all the principles, techniques, and tools in the MDSE galaxy, explaining the relationships and complementarities among them so that you can pick the ones that fit best your particular development scenario.

To complete the big picture the book also includes chapters on methodological and management aspects of MDSE projects to maximize your chances of success when starting the MDSE path. As with any other big change, it's not going to be easy but we do believe it's going to be worthwhile and we hope this book will provide you with the keys to succeed.

Bibliography

[1] Roberto Acerbis, Aldo Bongio, Marco Brambilla, Massimo Tisi, Stefano Ceri, and Emanuele Tosetti. Developing eBusiness solutions with a model driven approach: The case of Acer EMEA. In Luciano Baresi, Piero Fraternali, and Geert-Jan Houben, Eds., *Proc. of the 7th International Conference Web Engineering (ICWE'07)*, volume 4607 of *Lecture Notes in Computer Science*, pages 539–544. Springer, 2007. DOI: 10.1007/978-3-540-73597-7. 2

[2] Alfred V. Aho, Monica S. Lam, Ravi Sethi, and Jeffrey D. Ullman. *Compilers: Principles, Techniques, and Tools*, 2nd ed., Addison-Wesley, 2007. 141

[3] Manoli Albert, Jordi Cabot, Cristina Gómez, and Vicente Pelechano. Automatic generation of basic behavior schemas from UML class diagrams. *Software and System Modelling*, 9(1):47–67, 2010. DOI: 10.1007/s10270-008-0108-x. 27

[4] Scott Ambler. *Agile Modelling: Effective Practices for eXtreme Programming and the Unified Process*, Wiley, 2002. 56

[5] Sven Apel and Christian Kästner. An overview of feature-oriented software development. *Journal of Object Technology*, 8(5):49–84, 2009. DOI: 10.5381/jot.2009.8.5.c5. 60

[6] Dinesh Batra and George M. Marakas. Conceptual data modelling in theory and practice. *European Journal on Information Systems*, 4(3):185–193, 1995. DOI: 10.1057/ejis.1995.21. 21

[7] Benoit Baudry, Sudipto Ghosh, Franck Fleurey, Robert B. France, Yves Le Traon, and Jean-Marie Mottu. Barriers to systematic model transformation testing. *Communications of the ACM*, 53(6):139–143, 2010. DOI: 10.1145/1743546.1743583. 171

[8] Kent Beck. *Test-Driven Development by Example*, Addison-Wesley Professional, 2002. 58

[9] Alex E. Bell. Death by UML fever. *Queue*, 2(1):72–80, 2004. DOI: 10.1145/984458.984495. 74

[10] Jean Bézivin. On the unification power of models. *Software and System Modelling*, 4(2):171–188, 2005. DOI: 10.1007/s10270-005-0079-0. 8

[11] Mike Blechar and David Norton. Trends in model-driven development, 4Q09-3Q10. Technical report, Gartner, 2009. 3

[12] Petra Brosch, Gerti Kappel, Philip Langer, Martina Seidl, Konrad Wieland, and Manuel Wimmer. An introduction to model versioning. In Marco Bernardo, Vittorio Cortellessa, and Alfonso Pierantonio, Eds., *Advanced Lectures of 12th International School on Formal Methods for the Design of Computer, Communication, and Software Systems—Formal Methods for Model-Driven Engineering (SFM'12)*, volume 7320 of *Lecture Notes in Computer Science*, pages 336–398. Springer, 2012. DOI: 10.1007/978-3-642-30982-3. 161, 163

[13] Petra Brosch, Philip Langer, Martina Seidl, Konrad Wieland, Manuel Wimmer, and Gerti Kappel. The past, present, and future of model versioning. In *Emerging Technologies for the Evolution and Maintenance of Software Models*, pages 410–443. IGI Global, 2011. DOI: 10.4018/978-1-61350-438-3.ch015. 163

[14] Marco Brambilla, Piero Fraternali. Large-scale model-driven engineering of web user interaction: The WebML and WebRatio experience. *Science of Computer Programming*, 89:71–87, 2014. DOI: 10.1016/j.scico.2013.03.010. 115

[15] Hugo Bruneliere, Jordi Cabot, Cauê Clasen, Frédéric Jouault, and Jean Bézivin. Towards model driven tool interoperability: Bridging eclipse and microsoft modelling tools. In Thomas Kühne, Bran Selic, Marie-Pierre Gervais, and François Terrier, Eds., *Proc. of the 6th European Conference on Modelling Foundations and Applications (ECMFA'10)*, volume 6138 of *Lecture Notes in Computer Science*, pages 32–47. Springer, 2010. DOI: 10.1007/978-3-642-13595-8. 34

[16] Richard M. Burton and Borge Obel. *Strategic Organizational Diagnosis and Design*, Kluwer Academic Publishers, 1998. DOI: 10.1007/978-1-4684-0023-6. 54

[17] Antonio Cicchetti, Davide Di Ruscio, Romina Eramo, and Alfonso Pierantonio. Automating co-evolution in model-driven engineering. In *Proc. of the 12th International IEEE Enterprise Distributed Object Computing Conference (EDOC'08)*, pages 222–231, IEEE Computer Society, 2008. DOI: 10.1109/edoc.2008.44. 165

[18] Jim Conallen. *Building Web Applications with UML*, Addison-Wesley, 2002. 116

[19] Krzysztof Czarnecki and Simon Helsen. Feature-based survey of model transformation approaches. *IBM Systems Journal*, 45(3):621–645, 2006. DOI: 10.1147/sj.453.0621. 123

[20] Brian Dobing and Jeffrey Parsons. How UML is used. *Communications of the ACM*, 49:109–113, 2006. DOI: 10.1145/1125944.1125949. 21

[21] Hartmut Ehrig, Karsten Ehrig, Ulrike Prange, and Gabriele Taentzer. *Fundamentals of Algebraic Graph Transformation*, Springer, 2006. DOI: 10.1007/3-540-31188-2. 132

[22] Eric Evans. *Domain-driven Design: Tackling Complexity in the Heart of Software*, Addison-Wesley Professional, 2004. 57

[23] Jean-Marie Favre and Tam Nguyen. Towards a megamodel to model software evolution through transformations. *Electronic Notes in Theoretical Computer Science*, 127(3):59–74, 2005. DOI: 10.1016/j.entcs.2004.08.034. 167

[24] Andrew Forward and Timothy C. Lethbridge. Problems and opportunities for model-centric vs. code-centric development: A survey of software professionals. In *Proc. of the International Workshop on Models in Software Engineering (MiSE'08) and (ICSE'08)*, pages 27–32, ACM, 2008. DOI: 10.1145/1370731.1370738. 21

[25] Martin Fowler. *UML Distilled: A Brief Guide to the Standard Object Modelling Language*, 3rd ed., Addison-Wesley, 2003. 23

[26] Mirco Franzago and Henry Muccini and Ivano Malavolta. Towards a collaborative framework for the design and development of data-intensive mobile applications In *Proc. of the 1st International Conference on Mobile Software Engineering and Systems (MobileSoft2014)*, pages 58–61, ACM, 2014. DOI: 10.1145/2593902.2593917. 116

[27] Erich Gamma, Richard Helm, Ralph Johnson, and John Vlissides. *Design Patterns: Elements of Reusable Object-Oriented Software*, Addison-Wesley Professional, Boston, 1995. 68, 87, 145

[28] Anne Geraci. *IEEE Standard Computer Dictionary: Compilation of IEEE Standard Computer Glossaries*. The Institute of Electrical and Electronics Engineers Inc., 1991. 33

[29] Robert L. Glass. Loyal opposition—frequently forgotten fundamental facts about software engineering. *IEEE Software*, 18(3):112–111, 2001. DOI: 10.1109/ms.2001.922739. 23

[30] Jack Greenfield, Keith Short, Steve Cook, and Stuart Kent. *Software Factories: Assembling Applications with Patterns, Models, Frameworks, and Tools*, John Wiley & Sons, 2004. DOI: 10.1007/978-3-540-28630-1_19. 61

[31] Jonathan Grudin. Groupware and social dynamics: Eight challenges for developers. *Communications of the ACM*, 37(1):92–105, 1994. DOI: 10.1145/175222.175230. 54

[32] Reiko Heckel, Jochen Malte Küster, and Gabriele Taentzer. Confluence of typed attributed graph transformation systems. In Andrea Corradini, Hartmut Ehrig, Hans-Jörg Kreowski, and Grzegorz Rozenberg, Eds., *Proc. of the 1st International Conference on Graph Transformations (ICGT'02)*, volume 2505 of *Lecture Notes in Computer Science*, pages 161–176. Springer, 2002. 136

[33] Zhenjiang Hu, Andy Schürr, Perdita Stevens, and James F. Terwilliger. Dagstuhl seminar on bidirectional transformations (BX). *SIGMOD Record*, 40(1):35–39, 2011. DOI: 10.1145/2007206.2007217. 139

[34] John Hutchinson, Mark Rouncefield, and Jon Whittle. Model-driven engineering practices in industry: Social, organizational and managerial factors that lead to success or failure. *Science of Computer Programming*, 89: 144–161, 2014. DOI: 10.1016/j.scico.2013.03.017. 21, 23, 53

[35] ISO/IEC. 14977:1996(E) Information technology—syntactic metalanguage—extended BNF, International Standard, 1996. DOI: 10.3403/01300022. 85

[36] Javier Luis Cánovas Izquierdo, Frédéric Jouault, Jordi Cabot, and Jesús García Molina. API2MoL: Automating the building of bridges between APIs and model-driven engineering. *Information and Software Technology*, 54(3):257–273, 2012. DOI: 10.1016/j.infsof.2011.09.006. 34

[37] Ivar Jacobson, Grady Booch, and James Rumbaugh. *The Unified Software Development Process*, 1st ed., Addison-Wesley Professional, 1999. 55

[38] Jean-Marc Jézéquel. Model-driven engineering for software product lines. *ISRN Software Engineering*, 2012:1–24, 2012 DOI: 10.5402/2012/670803. 60

[39] Cliff B. Jones, Peter W. O'Hearn, and Jim Woodcock. Verified software: A grand challenge. *IEEE Computer*, 39(4):93–95, 2006. DOI: 10.1109/mc.2006.145. 169

[40] Frédéric Jouault. Loosely coupled traceability for ATL. In *Proc. of the International Workshop on Traceability (ECMDA'05)*, pages 29–37, 2005. 131

[41] Frédéric Jouault, Freddy Allilaire, Jean Bézivin, and Ivan Kurtev. ATL: A model transformation tool. *Science of Computer Programming*, 72(1-2):31–39, 2008. DOI: 10.1016/j.scico.2007.08.002. 125

[42] Frédéric Jouault and Massimo Tisi. Towards incremental execution of ATL transformations. In Laurence Tratt and Martin Gogolla, Eds., *Proc. of the 3rd International Conference on Theory and Practice of Model Transformations (ICMT'10)*, volume 6142 of *Lecture Notes in Computer Science*, pages 123–137, Springer, 2010. 139

[43] Gabor Karsai, Holger Krahn, Claas Pinkernell, Bernhard Rumpe, Martin Schindler, and Steven Völkel. Design guidelines for domain specific languages. In *Proc. of the 9th OOPSLA Workshop on Domain-Specific Modeling (DSM'09)*, 2009. 92

[44] Steven Kelly and Risto Pohjonen. Worst practices for domain-specific modelling. *IEEE Software*, 26:22–29, 2009. DOI: 10.1109/ms.2009.109. 77

[45] Anneke Kleppe. *Software Language Engineering: Creating Domain-Specific Languages Using Metamodels*, 1st ed., Addison-Wesley Professional, 2008. 87

[46] Dimitrios S. Kolovos, Davide Di Ruscio, Alfonso Pierantonio, and Richard F. Paige. Different models for model matching: An analysis of approaches to support model differencing. In *Proc. of the International Workshop on Comparison and Versioning of Software Models (CVSM'09) and (ICSE'09)*, pages 1–6, IEEE Computer Society, 2009. DOI: 10.1109/cvsm.2009.5071714. 161

[47] Thomas Kühne. Matters of (meta-)modeling. *Software and System Modelling*, 5(4):369–385, 2006. DOI: 10.1007/s10270-006-0017-9. 86, 91

[48] Stephen J. Mellor and Marc J. Balcer. *Executable UML: A Foundation for Model-Driven Architecture*, Addison-Wesley, 2002. 27

[49] Tom Mens and Pieter Van Gorp. A taxonomy of model transformation. *Electronic Notes in Theoretical Computer Science*, 152:125–142, 2006. DOI: 10.1016/j.entcs.2005.10.021. 123, 124

[50] Bart Meyers and Hans Vangheluwe. A framework for evolution of modelling languages. *Science of Computer Programming*, 76(12):1223–1246, 2011. DOI: 10.1016/j.scico.2011.01.002. 165

[51] Daniel L. Moody. The "physics" of notations: Toward a scientific basis for constructing visual notations in software engineering. *IEEE Transactions on Software Engineering*, 35(6):756–779, 2009. DOI: 10.1109/tse.2009.67. 103

[52] Tien N. Nguyen, Ethan V. Munson, John T. Boyland, and Cheng Thao. An infrastructure for development of object-oriented, multi-level configuration management services. In *Proc. of the 27th International Conference on Software Engineering (ICSE'05)*, pages 215–224, ACM, 2005. DOI: 10.1109/icse.2005.1553564. 163

[53] Antoni Olivé. On the role of conceptual schemas in information systems development. In Albert Llamosí and Alfred Strohmeier, Eds., *Proc. of the 9th Ada-Europe International Conference on Reliable Software Technologies (Ada-Europe'04)*, volume 3063 of *Lecture Notes in Computer Science*, pages 16–34, Springer, 2004. 2

[54] Object Management Group (OMG). Architecture-driven modernization task force, 2007. http://adm.omg.org 50

[55] Object Management Group (OMG). MDA guide revision 2.0, 2014. http://www.omg.org/cgi-bin/doc?ormsc/14-06-01 43

[56] Object Management Group (OMG). Meta object facility, version 2.0, 2006. http://www.omg.org/spec/MOF/2.0/PDF 89

[57] Object Management Group (OMG). Unified modelling language superstructure specification, version 2.1.2, 2007. `http://www.omg.org/spec/UML/2.1.2/Superstructure`
`/PDF` 102

[58] Klaus Pohl, Günter Böckle, and Frank J. van der Linden. *Software Product Line Engineering:
Foundations, Principles, and Techniques*, Springer, 2005. DOI: 10.1007/3-540-28901-1. 59

[59] Louis M. Rose, Markus Herrmannsdoerfer, James R. Williams, Dimitrios S. Kolovos, Kelly
Garcés, Richard F. Paige, and Fiona A. C. Polack. A Comparison of Model Migration
Tools. In Dorina C. Petriu, Nicolas Rouquette, and Øystein Haugen, Eds., *Proc. of the
13th International Conference on Model Driven Engineering Languages and Systems (MOD-
ELS'10)*, volume 6394 of *Lecture Notes in Computer Science*, pages 61–75, Springer, 2010.
DOI: 10.1007/978-3-642-16145-2. 166

[60] Andy Schürr. Specification of graph translators with triple graph grammars. In *Proc. of
the 20th International Workshop on Graph-Theoretic Concepts in Computer Science (WG'94)*,
volume 903 of *Lecture Notes in Computer Science*, pages 151–163, Springer, 1994. DOI:
10.1007/3-540-59071-4_45. 131

[61] Shane Sendall and Wojtek Kozaczynski. Model transformation: The heart and soul
of model-driven software development. *IEEE Software*, 20(5):42–45, 2003. DOI:
10.1109/ms.2003.1231150. 123

[62] Thomas Stahl, Markus Voelter, and Krzysztof Czarnecki. *Model-Driven Software Development: Technology, Engineering, Management*, John Wiley & Sons, 2006. 92

[63] Friedrich Steimann and Thomas Kühne. Coding for the code. *Queue*, 3:44–51, 2005. DOI:
10.1145/1113322.1113336. 23

[64] David Steinberg, Frank Budinsky, Marcelo Paternostro, and Ed Merks. *EMF: Eclipse Modeling Framework 2.0*, 2nd ed., Addison-Wesley Professional, 2009. 143

[65] Perdita Stevens. A Landscape of Bidirectional Model Transformations. In *Proc. of the
International Summer School on Generative and Transformational Techniques in Software En-
gineering II (GTTSE'07)*, volume 5235 of *Lecture Notes in Computer Science*, pages 408–424,
Springer, 2007. DOI: 10.1007/978-3-540-88643-3_10. 139

[66] Eugene Syriani and Hans Vangheluwe. De-/re-constructing model transformation languages. *ECEASST*, 29:1-14, 2010. DOI: 10.14279/tuj.eceasst.29.407. 136

[67] Massimo Tisi, Frédéric Jouault, Piero Fraternali, Stefano Ceri, and Jean Bézivin. On the
use of higher-order model transformations. In *Proc. of the 5th European Conference on Model
Driven Architecture—Foundations and Applications (ECMDA-FA'09)*, volume 5562 of *Lecture Notes in Computer Science*, pages 18–33, Springer, 2009. DOI: 10.1007/978-3-642-
02674-4_3. 138

[68] Massimo Tisi, Salvador Martínez, Frédéric Jouault, and Jordi Cabot. Refining models with rule-based model transformations. Research Report RR-7582, INRIA, 2011. 137, 139

[69] Albert Tort, Antoni Olivé, and Maria-Ribera Sancho. An approach to test-driven development of conceptual schemas. *Data & Knowledge Engineering*, 70(12):1088–1111, 2011. DOI: 10.1016/j.datak.2011.07.006. 59

[70] Edward Tsang. *Foundations of Constraint Satisfaction*, Academic Press, 1993. DOI: 10.1016/c2013-0-07627-x. 170

[71] Markus Völter. A catalog of patterns for program generation. In Kevlin Henney and Dietmar Schütz, editors, *Proc. of the 8th European Conference on Pattern Languages of Programs (EuroPLoP'03)*, pages 285–320, 2003. 28

[72] Andrés Yie and Dennis Wagelaar. Advanced traceability for ATL. In *Proc. of the 1st International Workshop on Model Transformation with ATL (MtATL 2009)*, 2009. 131

Authors' Biographies

MARCO BRAMBILLA

Marco Brambilla is currently professor of Software Engineering and Web Science at Politecnico di Milano (Italy). He is a shareholder and scientific advisor at WebRatio, the company that produces the MDD tool WebRatio `http://www.webratio.com`, based on IFML, a domain-specific language for UI modeling standardized by the OMG. He is the founder of Fluxedo `http://www.fluxedo.com`. In addition, he is one of the inventors of IFML and WebML languages. His interests include conceptual models, tools, and methods for development of Web and mobile apps, Web services, crowdsourcing, user interaction, business processes, web science, big data, social media, and social content analysis. He has been visiting researcher at Cisco Systems (San José, CA) and at UCSD (University of California, San Diego). He has been visiting professor at Dauphine University, Paris. He is coauthor of *Designing Data-Intensive Web Applications* (Morgan-Kauffman, 2003), *Interaction Flow Modeling Language: Model-Driven UI Engineering of Web and Mobile Apps with IFML* (Morgan-Kauffman and OMG Press, 2014), several other books, and more than 100 scientific articles published in conferences and journals. He is active in teaching and industrial consulting projects on MDSE, Domain-specific Languages (DSL), BPM, SOA, and enterprise architectures.

Personal blog: `http://www.marco-brambilla.com/blog`

Home page: `http://home.dei.polimi.it/mbrambil`

JORDI CABOT

Jordi Cabot is an ICREA Research Professor at Internet Interdisciplinary Institute, a research center of the Open University of Catalonia (UOC), where he is leading the SOM Research Lab. Previously, he has been a faculty member at the École des Mines de Nantes and team leader of the AtlanMod Research group on an Inria International Chair and worked in Barcelona (Technical University of Catalonia), Toronto (University of Toronto), and Milan (Politecnico di Milano). His research falls into the broad area of systems and software engineering, especially promoting the rigorous use of software models and engineering principles in all software engineering tasks while keeping an eye on the most unpredictable element in any project: the people involved in it.

Blog on software development: `http://modeling-languages.com`
Home page: `http://modeling-languages.com`

MANUEL WIMMER

Manuel Wimmer is a professor in the Business Informatics Group (BIG) at TU Wien (Austria). He has been a visiting professor at TU München (Germany) and at the Philipps-University Marburg (Germany) as well as a research associate at the University of Málaga (Spain). He is involved in several research projects in the area of model-driven engineering, model versioning, and model transformation which resulted in numerous scientific publications in international journals and conferences. Since 2005, he has taught the model-driven engineering course (about 200 students each year) at TU Wien and holds several industry seminars on how to efficiently apply model-driven engineering in practice.

Home page: `http://www.big.tuwien.ac.at/staff/mwimmer`

Index